The Reputation Handbook

Surviving and Thriving in the Age of Hyper-Transparency

Andrea Bonime-Blanc

GEC Risk Advisory LLC

www.GECRisk.com

First published in 2014 by Dō Sustainability
87 Lonsdale Road, Oxford OX2 7ET, UK

Copyright © 2014 Andrea Bonime-Blanc

The moral right of the author has been asserted.

All rights reserved. No part of this publication may be reproduced, stored in a retrieval system, or transmitted, in any form or by any means, electronic, mechanical, photocopying, recording or otherwise, except as expressly permitted by law, without the prior, written permission of the publisher.

ISBN 978-1-910174-31-9 (eBook-ePub)
ISBN 978-1-910174-32-6 (eBook-PDF)
ISBN 978-1-910174-30-2 (Paperback)

A catalogue record for this title is available from the British Library.

Dō Sustainability strives for net positive social and environmental impact. See our sustainability policy at **www.dosustainability.com**.

Page design and typesetting by Alison Rayner
Cover by Becky Chilcott

For further information on Dō Sustainability, visit our website:
www.dosustainability.com

DōShorts

Dō Sustainability is the publisher of **DōShorts**: short, high-value ebooks that distil sustainability best practice and business insights for busy, results-driven professionals. Each DōShort can be read in 90 minutes.

New and forthcoming DōShorts – stay up to date

We publish 3 to 5 new DōShorts each month. The best way to keep up to date? Sign up to our short, monthly newsletter. Go to www.dosustainability.com/newsletter to sign up to the Dō Newsletter. Some of our latest and forthcoming titles include:

- *Management Systems for Sustainability: How to Successfully Connect Strategy and Action*   Phil Cumming
- *Understanding Integrated Reporting: The Concise Guide to Integrated Thinking and the Future of Corporate Reporting*   Carol Adams
- *Corporate Sustainability in India: A Practical Guide for Multinationals*   Caroline Twigg
- *Networks for Sustainability: Harnessing People Power to Deliver Your Goals*   Sarah Holloway
- *Making Sustainability Matter: How To Make Materiality Drive Profit, Strategy and Communications*   Dwayne Baraka
- *Creating a Sustainable Brand: A Guide to Growing the Sustainability Top Line*   Henk Campher
- *Cultivating System Change: A Practitioner's Companion*   Anna Birney
- *How Much Energy Does Your Building Use*   Liz Reason

- *Lobbying for Good: How Business Advocacy Can Accelerate the Delivery of a Sustainable Economy*   Paul Monaghan & Philip Monaghan
- *Creating Employee Champions: How to Drive Business Success Through Sustainability Engagement Training*   Joanna M. Sullivan
- *Smart Engagement: Why, What, Who and How*   John Aston & Alan Knight
- *How to Produce a Sustainability Report*   Kye Gbangbola & Nicole Lawler
- *Strategic Sustainable Procurement: An Overview of Law and Best Practice for the Public and Private Sectors*   Colleen Theron & Malcolm Dowden
- *Business Strategy for Water Challenges: From Risk to Opportunity*   Stuart Orr & Guy Pegram

Subscriptions

In addition to individual sales of our ebooks, we now offer subscriptions. Access 60+ ebooks for the price of 5 with a personal subscription to our full e-library. Institutional subscriptions are also available for your staff or students. Visit **www.dosustainability.com/books/subscriptions** or email **veruschka@dosustainability.com**

Write for us, or suggest a DōShort

Please visit **www.dosustainability.com** for our full publishing programme. If you don't find what you need, write for us! Or suggest a DōShort on our website. We look forward to hearing from you.

Abstract

REPUTATION RISK HAS BECOME STRATEGIC because of the age of hyper-transparency. Indeed, the reputation risk hit parade of the early twenty-first century all but confirms this. There are three types of consequences to reputational hits today, all of which require greater reputation risk awareness in the age of hyper-transparency, defined as big and small data traveling and impacting at the speed of light. There is a strong business case – both qualitative and quantitative – for effective reputation risk management. What is 'risk'? What is 'reputation'? What is 'reputation risk'? We examine a variety of reputational risks and examples. It is a different kind of risk and to tackle it properly its different nature needs to be gauged and understood. We examine a variety of characteristics of reputation risk that explain its special nature. Reputation risk is then placed within the risk universe; first, within the spectrum of possible risk management architecture an entity might have in place; and second within 'big buckets' of substantive types of risk. Reputation risk is different and cross-disciplinary. Reputation risk actors – those who do – and stakeholders – those with a stake in what they do – are then examined. An interesting phenomenon – the reputation risk 'Rashomon' effect – is discussed. The ownership responsibility of management and oversight responsibility of the board are outlined. A typology of possible reputation risk management strategies available to entities is presented based on the risk architecture maturity of the entity and the breadth and depth of reputation risk knowledge and awareness that may exist within the entity. A reputation risk management

ABSTRACT

toolkit of fifteen tools is presented as well as a summary of roles and responsibilities of key reputation risk management actors. Reputation risk management is a strategic imperative. It can occur more successfully in high integrity cultures with evolved and high integrity leaders but there are ways to optimize reputation risk management even in less evolved or more dysfunctional cultures. The future of effective reputation risk management points to not only risk mitigation but to the transformation of this risk into opportunity and value for the organization.

About the Author

DR ANDREA BONIME-BLANC is CEO and founder of GEC Risk Advisory LLC, a global governance, risk, integrity, and reputation advisory firm (www.GECRisk.com). She has two decades of senior executive experience leading governance, legal, ethics, compliance, risk, internal audit and corporate responsibility in private, public, established and start-up companies in technology, media, infrastructure, utility, defense, manufacturing and professional services (including Bertelsmann and PSEG). With over 20 years of board experience, she is currently a board chair, audit committee chair and chair emeritus. She started her career at Cleary Gottlieb and received a joint JD/PhD from Columbia University. She was born and raised in Europe, is fluent in Spanish, and has worked and traveled extensively throughout the world. Dr Bonime-Blanc was named a 2014 Top 100 Thought Leader in Trustworthy Business, writes the "Ask the GlobalEthicist" column for *Ethical Corporation Magazine*, and has written for *Foreign Affairs, The New York Times, El Pais, El Cronista* and *Expansion*. Her book *Spain's Transition to Democracy: The Politics of Constitution-making* (http://amzn.to/Z9bEXv), used in many universities around the world, was first published in 1987. She is adjunct professor at NYU and a frequent guest lecturer and keynote speaker at many universities, organizations and venues worldwide. She can be reached at **abonimeblanc@gecrisk.com**, @GlobalEthicist and gecrisk.com.

Acknowledgments

NO BOOK WRITES ITSELF and no author is an island. For these reasons, I want to extend heartfelt thanks and appreciation to close colleagues, friends, wise people and collaborators who have helped me to develop and complete this Handbook whether they actually knew they were doing so or not! First and foremost, I must thank three close friends and colleagues for their invaluable direct contributions to making this a better book: Jacqueline E. Brevard, my close collaborator, dear friend and confidant; Hans W. Decker, who is the personification of the emotionally intelligent CEO; and Emmanuel Lulin, one of the most innovative business ethics practitioners I know. I would also like to thank two mentors – my close friends and colleagues Professor S. Prakash Sethi and Professor Mary C. Gentile, two visionary thought leaders who have provided me with amazing friendship and guidance for a long time. Organizations and colleagues who have provided varied support and assistance on issues relevant to this Handbook include: The Conference Board (Marion Feigenbaum, Joseph Feuer, Salvatore Vitale); Llorente & Cuenca (Jose Antonio Llorente, Jorge Cachinero, Beatriz Herranz, Marta Fernandez); NYSE Governance Services (Erica Salmon Byrne, Eric Morehead, Jean-Marc Levy, Jennifer Mitchell); The Global Risk Institute in Financial Services (Michel Maila, Gregory Frank); Ethisphere (Tim Erblich, Kevin McCormack, Stefan Linssen); Afi & Fundacion Afi (Jose Antonio Herce, Emilio Ontiveros, Veronica Lopez); Corporate Excellence (Saida Garcia); Garrigues (Antonio Garrigues, Ricardo Gomez, Jesus de la Morena, Juan Pablo Regojo Balboa, Myriam Gomez Alonso); The Institute

ACKNOWLEDGEMENTS

for Ethical Leadership at Rutgers University (Judy Young, Alex Plinio, James Abruzzo, Margaret Della); The National Association for Corporate Directors (Reatha Clark King, Erin Essenmacher); Ethical Corporation Magazine (Zara Maung, Greg Cohen). Individuals whom I have counted on or admire for their work and wisdom include, in alphabetical order: Lee Augsburger, Silvina Bacigalupo, Yelena Barychev, Donna Boehme, Lyn Boxall, Lanny Breuer, Adina Cale, Ted Coine, Keith T. Darcy, Nichola Dyer, Lauren Ferrari, David Finnie, Philippa Foster-Back, Ken Frankel, Margarita Garcia, Tom Gibson, Jeffrey Gracer, Pat Harned, Ulrika Hasselgren, Jonathan Hemmus, Annette Heuser, Jeff Hoffmann, Jessica Jimenez, Raquel Jurado Tiscar, Jeff Kaplan, Patrick Kelly, Barbara Brooks Kimmel, Mark Kollar, Dr. Nir Kossovsky, Richard Lang, Richard LeBlanc, Ed Levy, John C. Lenzi, Claudia Maskin, Stacey Millett, Joseph Murphy, Gwladys Ngo Tedga, Matthew Pachman, Kenneth Powell, Steve Priest, Willem Punt, Zain Raheel, Sonsoles Rubio Reinoso, Rupal Sankalia, Kerry Sulkowicz, Lori Tansey Martens, Emmanuel Tchividjian, Davia Temin, Adam Turteltaub, Coenraad van Beek, Rebecca Walker, Frank Wander, Toby Webb, Andrew Weissmann, Ian Welsh, Mark Winston, Gretchen Winter, Jay Worenklein, David Zweighaft. I would also like to thank my global collaborators and partners especially Murray Grainger (Spain), Jay Rosenzweig (Canada), Alexandra Mihaelescu Cichon (Switzerland), Kathrin Niewiarra (Germany), Craig and Julian Dinsell (US & UK), Michael Hickman (China), Simon Franco (Brazil), Matthias Kleinhempel and Gabriel Cecchini (Argentina). I would like to acknowledge the amazing insights and warmth of my dear friend, Jude Callirgos Robinson, who supported my writing this book as she supported everything else I did, and who we lost way too soon earlier this year. A special thanks go to my editor at Dō Sustainability, Nick Bellorini, for asking me to write this book

THE REPUTATION RISK HANDBOOK:
SURVIVING AND THRIVING IN THE AGE OF HYPER-TRANSPARENCY

in the first place. A big shout out goes to my lifetime group of friends in Malaga, Spain, whom I've known since we were 6 and who, through the marvels of social media, have humorously supported me throughout the writing of this book. Finally, I cannot thank three people enough: my wonderful partner and husband for his wisdom, my cool son for his creative passion and my beautiful and resilient mom – thank you, thank you, thank you.

Andrea Bonime-Blanc
New York City, September 2014

Contents

Abstract 5

About the Author 7

Acknowledgments 9

Introduction 17
Purpose and organization of the handbook 17

I UNDERSTANDING REPUTATION RISK 21

1 Reputation Risk in the
  Age of Hyper-transparency 23
  Mae West and the age of hyper-transparency 23
  'Reputation' across the ages 24
  Twenty-first-century reputation risk hit parade 25
  Reputational hits and consequences 27
  The age of hyper-transparency 30
  **The business case for effective
  reputation risk management** 35

2 Dissecting 'Reputation Risk' 41
  What is 'risk'? 41
  What is 'reputation'? 42

CONTENTS

	What is 'reputation risk'?	43
	Reputation risk is an amplifier risk: examples	45
	Other special qualities of reputation risk	48

II TRIANGULATING REPUTATION RISK 53

3 Reputation Risk Within the Risk Universe 55
  Reputation risk and risk architecture 55
  Reputation risk in the pantheon of
  'big buckets' of risk 58
  Reputation risk is different 61

4 Reputation Risk Actors & Stakeholders 63
  Reputation risk and you 63
  Reputation risk and the organization:
  actors & stakeholders 65
  The reputation risk 'Rashomon' effect 69
  Reputation risk ownership and oversight 72

III DEPLOYING REPUTATION RISK 79

5 Reputation Risk Strategies and Toolkit 81
  Reputation risk management strategies:
  a typology 81
  The reputation risk management toolkit 84

6 Optimizing Reputation Risk Management:
  the Next Strategic Imperative 95
  How to optimize reputation risk management 95

Visualizing reputation risk management within the organization: the big picture ... 104

Beyond strategic: effective reputation risk management can be transformational ... 105

Conclusion: The Way Forward ... 109

Notes ... 111

Introduction

Purpose and organization of the handbook

REPUTATION MANAGEMENT has been around for a while. Reputation risk, however, is a relatively new, largely unexamined and still evolving topic, which could easily be the subject of a more elaborate or scholarly book. This is not that book. This Handbook is about providing organizational practitioners – managers, executives and directors – with context, content and tools to accomplish two core strategic objectives: understanding and managing reputation risk and transforming reputation risk into value.

Reputation risk has never been a broader, faster, longer, or tougher topic. It's broader because reputation risk lurks everywhere – it's ubiquitous. It's faster because it can hit in a millisecond. It's longer because its digital footprint will exist somewhere forever. And it's tougher because we live in a complex, fast changing, dizzying world where we must constantly gauge what's going on. This book is about reputation risk management in the age of hyper-transparency.

While this Handbook is mostly focused on the corporate world, our review and treatment of reputation risk applies to all types of entities: business (corporations, joint ventures, partnerships and associations), non-profit (local, international, big and small), government (country, regional or global), research and educational.

INTRODUCTION

The Handbook is written by a practitioner for practitioners, based on decades of front line reputation risk management by any other name: as a senior executive inside four global companies in diverse sectors, leading legal, governance, compliance, ethics, audit, risk, crisis, business continuity, corporate responsibility, strategic communication, information security, environmental, health and safety functions and initiatives; as a member and/or chair of multiple boards of directors and committees; as a member or leader of several global organizations and associations; as a speaker, teacher, columnist and author; and, most recently, as the CEO and founder of a strategic global governance, integrity, risk and reputation advisory firm. I have seen a full 360 degrees of reputation risk several times over.

While reputation management has been traditionally part of marketing, branding, public relations and communications, reputation risk is a broader, deeper, more multifaceted subject that requires additional and varied skills. In essence, everyone in an organization – from the board member and CEO to the administrative assistant and new employee – has a role to play in reputation risk management and owns at least a sliver of their organization's reputation risk portfolio. And, at the end of the day, each of us owns 100% of our personal reputation risk management.

So what is organizational reputation risk? It's the risk of something affecting an organization's good name with the potential to damage current or future prospects, potentially materially. Yet embedded in this concept is also an upside: a chance to enhance reputation. Therein lies the beauty of effective reputation risk management – it's not just about protecting the downside, it's about enhancing the upside.

**THE REPUTATION RISK HANDBOOK:
SURVIVING AND THRIVING IN THE AGE OF HYPER-TRANSPARENCY**

Handbook overview

The Handbook has three parts – Understanding, Triangulating and Deploying Reputation Risk – and six chapters as follows:

Part I – Understanding reputation risk. Provides context and introduces the elemental themes of reputation risk in two chapters:

- Chapter 1 – 'Reputation risk in the age of hyper-transparency' provides the context and big picture on why reputation risk has suddenly become more important in today's hyper-connected, big data-driven, fast changing world. It concludes by making the business case for effective reputation risk management.

- Chapter 2 – 'Dissecting "reputation risk"' breaks down the concept of 'reputation risk' through definitions, examples, and a review of some of its unique characteristics.

Part II – Triangulating reputation risk. Part II focuses on wrangling and taming reputation risk in two chapters:

- Chapter 3 – 'Reputation risk within the risk universe' frames reputation risk by 1) reviewing the risk management infrastructure within which it might exist; 2) asking whether it is one of the 'big bucket' categories of substantive risk that organizations typically face or something else; and 3) relating it to crisis management and organizational integrity.

- Chapter 4 – 'Reputation risk actors and stakeholders' reviews 1) the main organizational reputation actors; 2) the key internal and external stakeholders with a 'stake' in the entity's reputation; and 3) who bears organizational reputation risk ownership and oversight responsibility.

INTRODUCTION

Part III – Deploying reputation risk. The concluding two chapters outline practical options for a successful deployment of reputation risk management:

- Chapter 5 – 'Reputation risk strategies and toolkit' outlines 1) a typology of possible reputation risk management strategies; 2) a reputation risk management toolkit with fifteen components; and 3) guidance on the principal roles and responsibilities of key actors in deploying the various tools in the toolkit.

- Chapter 6 – 'Optimizing reputation risk: The next strategic imperative' looks at 1) optimizing reputation risk management strategies by understanding an organization's leadership style and culture; 2) where reputation risk management fits within the broader dynamics of an organization; and 3) the way forward on reputation risk – reputation opportunity and enhancement.

PART I
UNDERSTANDING REPUTATION RISK

CHAPTER 1

Reputation Risk in the Age of Hyper-transparency

I lost my reputation but, then again, I didn't need it.
MAE WEST, AMERICAN FILM ACTRESS, 1930S

Mae West and the age of hyper-transparency

MAE WEST, the lusty and irrepressible American movie actress of the early twentieth century was responsible for the above quote and this one: 'When I'm good I'm very good, but when I'm bad I'm even better.' For her, being bad was good for business. That was her reputation – to be bad. Losing her reputation wasn't the problem – but losing her reputation for 'being bad' was her greatest reputation risk and would have been a blow to her livelihood.

Can we apply this approach to today's business and organizational context?

Most organizations want to build and retain a 'good reputation' for whatever it is that they do or offer. It isn't necessarily about being good or bad but about consistently and predictably doing what you do best: creating products, providing services, creating some form of value.

However, behaving 'badly' in the new age of hyper-transparency can be hazardous to an organization's health. The damage can be instant, very public and, in some cases, irreversible. It's no longer just about making products, delivering services or creating value anymore; it's about doing these things under the extreme spotlight of a hyper-transparent world.

Extrapolating to organizational life, Mae West got it right, but only half right. Maintaining and improving your reputation – for whatever that might be – is different from being 'good' or 'bad'. While these words are simplistic, charged and relative, the point is this: in today's hyper-transparent world, organizations need to do both things – build and defend their reputations and be (or be perceived to be) 'good' in the eyes of most stakeholders. The recent annals of reputations lost and never recovered are littered with examples of companies that did neither: Enron, Lehman Brothers, Barings and WorldCom come to mind.

There's a reason why we have seen so many more of these cases since the turn of the century – the age of hyper-transparency has changed the very nature of 'reputation' from something somewhat amorphous and superficial to something more material and impactful. Indeed it may very well be that the age of hyper-transparency is the handmaiden of the relatively new and still misunderstood concept of 'reputation risk'.

'Reputation' across the ages

It's not that reputational matters are new – indeed reputation is an age-old concept. We can go back to the fourth century BC to find Socrates' wisdom on the subject:

> *Regard your good name as the richest jewel you can possibly be possessed of – for credit is like fire; when once you have kindled it you may easily preserve it, but if you once extinguish it, you*

> will find it an arduous task to rekindle it again. The way to a good reputation is to endeavor to be what you desire to appear.

A little later, in the first century BC, Publilius Syrus said:

> A good reputation is more valuable than money.

In the nineteenth century Abraham Lincoln stated:

> Character is like a tree and reputation like a shadow. The shadow is what we think of it; the tree is the real thing.

And, then of course, everyone knows the famous words of present-day business titan, Warren Buffet:

> It takes 20 years to build a reputation and five minutes to ruin it. If you think about that, you'll do things differently.

Buffet is also quoted as saying:

> We can afford to lose money – even a lot of money. We cannot afford to lose reputation – even a shred of reputation.[1]

Many wise sayings and much time traversed and yet the essence of the meaning of 'reputation' remains pretty much the same: 'Reputation' is about the perception by others (stakeholders) of the state of something (an entity, product or service) or someone (a leader or other person) and the danger (of loss) or opportunity (of gain) that such perception provides to such entity, thing or person.

Twenty-first-century reputation risk hit parade

Reputation hits can affect the largest entities in the world – including

REPUTATION RISK IN
THE AGE OF HYPER-TRANSPARENCY

the most powerful governments and global corporations. America's reputation, for example, has suffered a variety of blows over the past decade and a half. First, unpopular wars – Iraq and Afghanistan – and then National Security Agency (NSA) high-tech spying revelations on US citizens and friendly governments. Specific reputational damage has resulted not only to the US government but also, by association, to its citizens and even its technology sector which may now be less trusted by non-US stakeholders (customers) than before.[2]

The BP Deepwater Horizon oil spill disaster of 2010 was the biggest of its kind ever in terms of sheer magnitude and resulting attention, fines, settlements, civil and criminal investigations, and reputational damage. BP's then CEO, Tony Hayward, made things worse when he publicly complained about not having enough time off, seemingly putting his personal comfort ahead of a terrible crisis that had just caused 11 deaths and unprecedented environmental and financial damage.

Reputation hits can affect entire sectors, globally. The reputational hit parade of the past decade in the global financial sector has been non-stop and unparalleled, involving almost every big bank name – JP Morgan, RBS, HSBC, Standard Chartered, Goldman, Citibank, Barclays, UBS, Deutsche Bank, BNP Paribas, Credit Suisse and more. The banking industry seems to think of reputation risk, ethics and compliance, amorphously at best or as a 'cost of doing business'. It's not surprising that this sector places last consistently over time in industry sector trust surveys, like the Edelman Trust Barometer which has been gauging stakeholder trust in industry and government for over a decade. However, the tide may be turning as recent mega-fines, regulatory overdrive and stock under-performance may be starting to put a dent in what previously seemed to be a sector that didn't seem to notice the importance of reputational risk.[3]

And then there is the slow-motion, long-term, reputational unraveling. After surviving a major existential threat, going bankrupt and being saved by a massive government bailout, GM seems to have done it again. In January 2014, an investigation revealed that thirteen GM car accident deaths (and many more injuries) occurring over the past decade appeared to be the result of a defective ignition switch that could have been fixed for $1 per car. Apparently for cost-saving reasons (traced back to deep-seated cultural dysfunction), the correction was never made, the problem wasn't disclosed and a cover-up ensued. The full magnitude of this reputational hit is yet to unfold but is unlikely to be modest or short-lived.[4]

Reputational hits and consequences

Does a reputational hit today mean long-term reputational damage? Or are these mostly momentary blips that affect certain stakeholders (investors, customers) temporarily but, depending on the response of the organization, won't lead to long-term negative consequences?

Measuring reputational risk is a challenging and as yet unconquered art or science but attempts are being made. For example, the impact of a specific event on a company's publicly traded stock can be a useful metric though obviously useless for privately held businesses and other organizational forms like NGOs or government agencies.

It is helpful, however, to look at stock metrics, as there are lessons to be learned. Wal-Mart stock declined by almost 5% (or US$10 billion) the trading day after the *New York Times* published its in-depth investigative

**REPUTATION RISK IN
THE AGE OF HYPER-TRANSPARENCY**

report on Wal-Mart's alleged corruption and bribery of Mexican officials on 20 April 2012.[5] Similarly, observers charted a decline of US$7 billion over four days of trading in NewsCorp stock when allegations of widespread phone-hacking by News of the World journalists were first reported in July 2011.[6]

However, in both cases, there have been other longer-term quantitative and qualitative hits: over 1000 employees of News of the World lost their jobs when the paper closed within a year of the revelations; high level executives and even members of the board from each company have been either terminated, investigated and/or prosecuted; hundreds of millions of dollars (and multiple thousands of hours of employee time) are still being spent annually by each company because of ongoing investigations, legal cases and internal compliance reforms.

There are essentially three possible types of outcome from a reputational hit depending on how prepared and effective an organization and its leaders are. There is the 'Deadly Blow' where the consequences of a serious reputational risk gone wrong are devastating to people, products, profits and/or performance. There is no recovering – witness Enron, Arthur Andersen, WorldCom, Lehman Brothers (see Table 1). There is the 'Recoverable Hit' where recovery and even regeneration are possible as in the case of the German corporate giant Siemens, which overcame its costly (US$2 billion+) enterprise-wide corruption scandal of the early 2000s and adopted what is considered the gold standard for global corporate anti-corruption programs. And then, though rare, there are reputation 'Enhancement Events' such as the case of Johnson & Johnson (J&J) in the mid-1980s and its masterful handling of a potentially disastrous event – the criminal tampering of Tylenol containers outside of the company's control resulting in several deaths. The company's quick, responsible and systematic action not only prevented further deaths but truly enhanced their reputation for responsibility towards stakeholders (especially customers).

TABLE 1. Reputational hits and consequences

Reputational hit type	Description	Example
Deadly blow	Organization, product, service &/or leader disappears	Enron, Lehmans, News of the World
Recoverable hit	Organization, product, service &/or leader regroups and recovers	Siemens, BP
Enhancement event	Organization, product, service &/or leader is prepared in advance & builds reputational equity over time	Johnson & Johnson

However, unlike J&J, most entities consider reputation risk in one of two ways – after the fact, as a result of a reputational hit, or grudgingly, in response to other serious unrelenting pressures. In the first instance, risk and crisis management don't really exist and the only possible response to a reputational hit is to scramble and gamble, with all the chaos and mismanagement that entails. Or, because of regulatory pressures or other stakeholder 'demands', some entities create an often under-resourced or superficial risk, compliance or ethics program, or a marketing-oriented corporate social responsibility (CSR) program

REPUTATION RISK IN THE AGE OF HYPER-TRANSPARENCY

that's supposed to demonstrate 'good' corporate citizenship (and 'good' reputation).

These efforts are skin-deep – a 'Potemkin Village' approach to reputation management generally with a pretty veneer but no substance from a risk or reputation risk management standpoint. Both of these responses are reactive and inadequate and are neither sustainable nor responsible, certainly not in the age of hyper-transparency.

Until the very recent past, executives and boards haven't considered reputation risk management at all or as important as focusing on more 'important' strategic business and financial risks. But this is changing and fairly dramatically as boards and executives suddenly realize that reputation risk is pervasive: in the last year, major surveys of global CEOs, executives and boards, singled out reputation risk as either one of the top or the number one strategic risk. What happened?[7]

The age of hyper-transparency

Transparency + data + velocity = hyper-transparency
(Big & small data traveling & impacting at the speed of light)

The age of hyper-transparency is what happened. The age of Wiki-Leaks and Edward Snowden's NSA revelations is what happened. The age of Apple iCloud hacked celebrity nude photos is what happened.

The age of hyper-transparency is best characterized by the words: 'Nowhere to run, nowhere to hide.' In the corporate world, this adage is well illustrated by the constantly evolving and complex GlaxoSmithKline (GSK) corruption scandal which began in China in 2013 with a government investigation into what first appeared to be corruption issues, and has since morphed and expanded into a multifaceted set of issues involving sex, privacy, lies and videotape. Because of the hyper-transparent age we live in, multiple governments, prosecutors, investigators, the media and others have jumped on the GSK investigatory bandwagon and are looking under every possible rock, including in such unlikely and unrelated places like Iraq and Syria.[8]

There are four factors that explain why reputation risk is suddenly considered important and even strategic:

- Scholarly and practitioner work on the real value of intangible things.[9]

- The proliferation and visibility of high impact scandals since the turn of the century.

- Growing awareness and impact of non-financial considerations: namely, environmental, social and governance (ESG).

- Greater transparency, availability, velocity and volume of information through technology.

Finally, this age of hyper-transparency also includes the proliferation of a multitude of players equipped or claiming to be equipped to be reputation experts. Let's take a quick look at a few of these experts.

Reputation educators

Educational institutions and think tanks are offering more courses on reputation management, though catering mostly to the public relations and communications field. Increasingly, however, there are scholarly inquiries into the nature of reputation and reputation risk, with think-tanks that study, analyze and publish important contributions to the scholarly and practical dialogue. Corporate Excellence: The Centre for Reputation Leadership headquartered in Spain, for example, has recently published a major contribution to this topic.[10]

Reputation analysts

There are a few reputation measuring services and analysts; for example, the consultancy the Reputation Institute has a longstanding 'Global RepTrak' which rates and tracks the reputation of a variety of entities. One of the few and more interesting entries in this field, focused exclusively on reputation risk analytics, is Swiss-based RepRisk ESG Business Intelligence. They deploy big data analytics and algorithms to dynamically analyze and quantify reputational risk and assign a RepRisk Index (RRI) to companies, sectors, countries, issues and themes based on extra-financial, ESG risks. The RRI 'is a risk measure that captures criticism and quantifies exposure to controversial ESG issues, such as environmental degradation, human rights abuses, corruption and more. It does not measure a company or project's overall reputation, but rather is an indicator of their reputational risk.'[11] See Figure 1 for an example of the RRI showing a snapshot of most exposed companies from August 2012 through August 2014.

FIGURE 1. RepRiskIndex (RRI): List of most exposed companies related to ESG issues (August 2012–August 2014).

Peak RRI 75-100

Name	RRI Current / Change / Peak	Sectors	Location
Chonghaejin Marine Co Ltd	57	-12	93	Travel and Leisure	Korea, Republic of (South Korea)
Chang Guann Co Ltd	86	+86	86	Food and Beverage	Taiwan
Kunshan Zhongrong Metal Products Co	75	-8	83	Automobiles and Parts	China
Tazreen Fashions Ltd	32	-1	80	Personal and Household Goods	Bangladesh
Federation Internationale de Football Association (FIFA)	61	-13	79	Travel and Leisure	Switzerland
Dongguan ShinYang Electronics Ltd	72	+3	77	Electronic and Electrical Equipment	China
General Motors Co (GM)	75	+0	75	Automobiles and Parts	United States of America
Shanghai Husi Food Co Ltd (Shanghai Fuxi Food)	70	+0	75	Food and Beverage	China

Peak RRI 50-75

Name	RRI Current / Change / Peak	Sectors	Location
KT ENS Corp	31	-2	74	Software and Computer Services	Korea, Republic of (South Korea)
Liberty Reserve	24	-3	74	Financial Services	Costa Rica

SOURCE: Reprinted with permission from RepRisk. © RepRisk ESG Business Intelligence 2014. All Rights Reserved. www.reprisk.com

Reputation consultants

Reputation consultants, traditional and non-traditional, are proliferating, ranging from large global firms (including, recently, accounting 'Big Four' and traditional management consultancies) to specialized boutiques. Most of these firms come from the communications and public relations field but increasingly are looking at reputation risk as well and beefing up their capabilities with multi-disciplinary teams.

Reputation insurers

Reputation risk insurance has also developed recently, from large companies like AIG and Zurich to specialized outfits like Steel City Re. Dr Nir Kossovsky, a pioneer in the reputation risk and reputation risk insurance space and CEO of Steel City Re, has written extensively on this subject and explains reputation risk insurance as follows:

> *Reputation insurance coverage. . . belong(s) to the same class of value-added service coverages as kidnap and ransom insurance. . . where the actuarial value of the covered losses (premium/burden vs frequency/severity) is secondary. More important are the expert risk mitigation and risk management services embedded in the policy-based solutions. These services prevent issues rather than just compensate for the costs of problems after the fact. These policies can provide risk mitigation before, during and after an incident through controls and preparedness, including crisis communications and messaging support and media training to help retain stakeholder loyalty and protect enterprise value.*[12]

Reputation 'protectors'

A number of services – most of them online and virtual – have sprung up in this age of hyper-transparency to cater to the needs, desires and, yes, narcissism of the modern age. The need to scrub one's online identity has become de rigueur for those in the limelight, those falsely (or accurately) accused of improprieties or worse, the very wealthy or famous who want to be virtually 'absent' and corporations and other organizations seeking to control their image and brand (reputation.com is an example of this kind of service).

Reputation raters

The age of hyper-transparency has brought with it an onslaught of rankings and ratings of all kinds from a wide variety of media and non-media outlets. These range in quality, standards and rigor but all touch upon the reputation (or reputation risk) of what they are targeting. These range from the well-regarded and rigorous Harris Reputation Quotient, to very basic, unsophisticated, crowd-sourced, online surveys like 'Glassdoor' or 'Rate My Professor'.

In between are a vast variety of somewhat rigorous but mostly marketing oriented rankings like the Forbes Best Places to Work, a multitude of similar work/life, diversity and other company rankings and a multitude of college and university rankings, for example.

Finally, there are the well-known government or NGO-based ratings that also touch on reputational issues: one of the most widely respected and used is Transparency International's annual Corruption Perceptions Index which rates countries on how corrupt they are perceived to be.

The business case for effective reputation risk management

The case for effective reputation risk management in this age of hyper-transparency can be made in two ways – accentuating the positive and exposing the negative. There is growing quantitative and qualitative evidence that smart reputation risk management can add value to the bottom line – through liability avoidance, cleaner and leaner processes and improved products and services. Indeed, properly deployed and integrated, effective reputation risk management can be transformational, actually adding value to the financial bottom line.

REPUTATION RISK IN THE AGE OF HYPER-TRANSPARENCY

The financial sector has certainly been in the eye of this storm in recent years given the massive impact of questionable, illegal and downright criminal behaviours exhibited in this sector. Indeed, there is a growing body of work that is zeroing in on linking ethical behaviours (and their absence) and the attendant culture, to results and performance.[13]

A tack some of the 'too big to fail' players have recently taken is to throw vast amounts of money (billions of dollars) and people (thousands) at their 'compliance' problem. They are, however, missing the critical point here. Theirs is not a 'compliance' problem, theirs is a systemic cultural and leadership problem in need of a more strategic solution. It's not enough to throw vast resources at a systemic problem. The real challenge begins and ends at the very top: the CEO and the board. Investors are also beginning to understand this as the still nascent but growing activist and institutional investor movement for greater transparency has recently gathered steam.

Whether as lip service to the regulator, or because their bottom lines are starting to feel the pinch, some of the global banks may be getting it. Some are appointing high-level chief compliance officers who are part of the executive team, have a seat at the table and/or access to the board (HSBC, JP Morgan). Others have taken a different approach. For example, after its troubles in 2010, Goldman created a high level committee, issued thirty-nine business standard principles and started a massive reputation management effort. Yet others are creating sustainable investment, community engagement programs and 'green' measures to demonstrate their reformist bona fides (Morgan Stanley, Goldman). Smart reputation risk management is part of a long-term strategy, not a one-time fix. While the jury is out on these recent developments they provide hope.

The quantitative case: The growing evidence

One of various longitudinal studies that have been done tracked 180 publicly traded companies over eighteen years and found that those with sustainability programs performed substantially better on three different financial metrics (see Table 2).[14]

TABLE 2. Long-term positive financial impact of sustainability

	An investment in 1993 of:	Results in 2010 for sustainable companies	Results in 2010 for low/no sustainability companies
Investment	$1	$22.6	$15.4
Return on equity	$1	$31.7	$25.7
Return on assets	$1	$7.1	$4.4

Source: 'The impact of a corporate culture of sustainability on corporate behavior and performance', by Robert G. Eccles, Ioannis Ioannou and George Serafeim. Working Paper, May 2012.

Dr Kossovsky, a practitioner in the reputation insurance field, has concluded that a 'reputational crisis will shave an average 7% from a company's market capitalization' and that reputational value restoration can yield a net additional average of 13.5% in market capitalization. Companies that have voluntarily embraced a sustainable business culture over many years 'significantly outperform their counterparts over the long-term, both in terms of stock market and accounting performance'.[15]

These data are compelling and point in an important direction: reputation hits hurt companies, and reputation enhancing-work helps companies, financially and often substantially so. If companies took this fact more seriously, they would realize that deploying an effective reputation risk management strategy could unlock hidden value that otherwise lies fallow in the silos of the organization.

The qualitative case: Stakeholder trust at the core

The qualitative case for effective reputation risk management can be made in two ways – one accentuating the negatives that will be eliminated or at least tamed if there is effective reputation risk management. The other is to accentuate the positive – all the good things that will happen to an organization if it has effective reputation risk management in place. Table 3 provides a summary of these points most of which revolve around one critically important theme: maintaining and improving stakeholder trust lies at the core of effective reputation risk management.

While quantitative snapshots such as stock price analysis can be helpful, the qualitative measures of how key stakeholders seize up an entity are just as important or perhaps even more so when it comes to understanding an entity's long-term resilience and ability to recover its reputational health.

And that is because trust is at the core. When key stakeholders lose trust in the present and future reliability, efficacy, transparency, growth, quality or other tangible value of an entity, its leaders, products or services, that's when the reputational hit has a tangible impact, whether or not it is easy to measure. Reputations will remain damaged for as long as actions are not taken to regain that trust.[16] Table 3 summarizes the business case for effective reputation risk management.

TABLE 3. The business case for effective reputation risk management

Factor	The negative case	The positive case
Stock or other value impact	Average 7% drop on public company stock (potentially similar financial impact on non-stock entities) (Kossovsky data)	Average 13.5% net appreciation of public company stock when restoring reputation (potentially similar financial impact on non-stock entities) (Kossovsky data)
Investor impact	Loss of quality investors	Attract quality investors
Third party impact	Worse third party terms	Better third party terms
Costs & expenses	Extra cost of liability, legal, compliance	Lower legal, liability or compliance costs
Time & resources	Loss of time, resources, creativity of workforce sucked into investigations & litigation	Time and work resources focused on developing & doing business, value creation
Investigations	One investigation will get you more, many more	Transparency will avoid or minimize investigations
Personal reputations	Reputations adversely, even irreparably damaged, especially executives and board members	Reputations intact

REPUTATION RISK IN THE AGE OF HYPER-TRANSPARENCY

Factor	The negative case	The positive case
Business/mission impact	Restructuring or demise of business or business lines	Resilience of business and business lines
Consumer impact	Consumer dissatisfaction and sales, volume, pricing losses	Consumer satisfaction, more sales, volume, premium pricing
Employee impact	Employee dissatisfaction, cultural malaise	Employee satisfaction, esprit de corps
New talent impact	Hits on recruiting new talent, loss of good talent, lost jobs	More jobs, attraction & retention of coveted talent
Regulator impact	Bad to worse relationships with regulators in multiple locations	Regulators more forgiving when/if a problem occurs
Media impact	Under the media microscope	Good media coverage, if any
Social media impact	The social media watch is on and uncontrollable, magnifies reputational damage	Social media coverage may be positive & useful to enhancing reputation

CHAPTER 2

Dissecting 'Reputation Risk'

What is 'risk'?

WHETHER IT IS THE PROFIT MOTIVE of a company, the education imperative of a university, social services at a government agency or the mission to feed the hungry of a non-profit, organizations require resiliency to achieve their primary mission. In order to achieve resiliency, every organization needs to know what its risks are and build, fortify and protect its resilience and hard-earned reputation from the many and complex risks of today's irreversibly hyper-transparent world.

So what is a 'risk'?

- *'Risque'* was the spelling used for 'risk' starting in 1622. The spelling was changed to 'risk' in 1655 in the *Oxford English Dictionary* and defined as: '(Exposure to) the possibility of loss, injury, or other adverse or unwelcome circumstance; a chance or situation involving such a possibility.'[17]

- The latest and possibly greatest definition of 'risk' comes from the recently adopted revised definition under ISO 31000: risk now is the 'effect of uncertainty on objectives'.[18]

In this revised definition lies a much-needed new lens on risk – that of opportunity, of something potentially positive embedded in every

DISSECTING 'REPUTATION RISK'

risk. 'Risk' – both internal and external – has had generally negative connotations, and been associated largely with danger, harm, destruction, and bad behaviors. By using generally neutral terms like 'uncertainty' and 'objectives' the new definition does not cast a judgmental slant on risk. Risk can be negative, neutral and, yes, even positive.

It is also important to mention the difference between internal risks over which an entity may have a certain amount of control (for example, building proper cyber-defenses, building a robust culture of integrity through a proper incentive program) and external risks over which there is limited or no control (a natural disaster affecting a supply chain or causing health, safety or security impacts).

What is 'reputation'?

If we go back to the Mae West concept of reputation, having a 'bad' reputation or a reputation for being 'bad' was good for business, at least for her. And there are clearly organizations that fancy themselves edgy, risk-taking or into high-risk products and services that in turn encourage risky behaviors or activities. Amongst such entities might be casinos, tobacco companies, skateboard or jet-ski manufacturers or bungee jumping facilities. But all of these companies would want to retain their 'good' reputation for making these products or delivering these services well and safely.

Mainstream dictionaries provide the following guidance on 'reputation':

> ... *the beliefs or opinions that are generally held about someone or something [or] the widespread belief that someone or something has a particular habit of characteristic.*[19]

. . . the common opinion that people have about someone or something: the way in which people think of someone or something more fully.[20]

Reputations are about the perceptions, opinions and actions of others – stakeholders –with relation to the entity, product, service or person. Stakeholders play a critical role in all things reputational. Whether they are internal stakeholders (i.e. employees, shareholders) or external (i.e. customers, regulators), reputation risk is all about stakeholders' perception of an entity, its people, its products or services.

When a company with a troubled financial history, gets a new leader who turns things around in a systematic, thoughtful and constructive way, it is a positive reputation risk event. This is what happened when Alan Mullaly took over as Ford CEO taking the then very troubled auto company to one of the most successful turnarounds in corporate history. Quite the opposite seems to be happening at the other leading US auto manufacturer, GM. The negative reputation hit of the faulty ignition switch scandal is likely to continue to have a material and long-term impact on the company's reputation with its stakeholders, especially customers.

What is 'reputation risk'?

Let's start with a typology developed by Jorge Cachinero, an author and reputation management practitioner. He has developed the following classification of 'reputational risk', depending on where the risks are coming from:

1. Natural risks: Those determined by the natural environment including climatic, atmospheric or seismic events or phenomena that cannot be easily predicted by companies.

DISSECTING 'REPUTATION RISK'

2. Leadership risks: Those directly related to mistakes by organizations, their senior managers, in exercising their responsibilities as leaders.

3. Operational risks: Those arising as a result of the production process of each business, including aspects of the value chain, the supply chain and logistics.

4. Environmental risks: Important regulatory or legislative changes with a decisive effect on the operating environment of a specific industry or sector.[21]

TABLE 4. Classification of reputation risk

Type	Example	Risk Issues
Natural	TEPCO Fukushima event (2011)	Earthquake, tidal wave, safety
Leadership	Wal-Mart Mexico business decision (early 2000s to date)	Alleged bribery & corruption
Operational	Toyota supply chain troubles (2010)	Supply chain & defective parts
Environmental/ regulatory	Spanish banking system changes (since 2008)	Fraud, corruption, theft

Source: Jorge Cachinero, 'Reputational risks', forthcoming paper, 2015.

As the examples in Table 4 show, reputation risk is a broad and malleable concept and, because of this, presents challenges for a precise definition. However, it is this very malleability and amorphousness that is at the heart of what reputation risk is.

**THE REPUTATION RISK HANDBOOK:
SURVIVING AND THRIVING IN THE AGE OF HYPER-TRANSPARENCY**

This is our description of 'reputation risk', a term that will continue to evolve and mutate but whose essence can be summarized as follows:

Reputation risk is an amplifier risk that layers on or attaches to other risks – especially ESG risks – adding negative or positive implications to the materiality, duration or expansion of the other risks on the affected organization, person, product or service.

Reputation risk is an amplifier risk: examples

Let's illustrate this description with some examples:

Reputation risk and supply chain risk:
the retail industry & Rana Plaza

The retail industry for decades sought out the cheapest labor in the world by contracting with third party suppliers in the least regulated, least safe and poorest countries. They knew that they had supply chain risk when they entered into these relationships.

Despite many years of serious incidents in various locations, the global retail industry found out about reputational risk the hard way when the human disaster of Rana Plaza occurred in Bangladesh in 2013 when a building, well known to be structurally unsound and unsafe, housing many third party suppliers, collapsed, killing more than 1100 workers.

The resulting reputational hit on the likes of the GAP, Wal-Mart, H&M and many others propelled them to do more than they had done before: they created industry reputation rebuilding efforts through the creation of the Bangladesh Accord and the Bangladesh Worker Safety Alliance.

Whether or not these efforts are truly successful, they represent an

DISSECTING 'REPUTATION RISK'

acknowledgment by these companies that reputation risk is a serious and potentially strategic risk associated with their supply chain risk.

Reputation risk and labor & human rights risk:
Apple & Foxconn

A similar story can be told about Apple's experience with Foxconn, their major outsourced supplier of labor in China where most of Apple's manufacturing takes place. When Apple contracted with Foxconn they probably planned for third party contractual risk. However, they didn't seem prepared to deal with the labor and human rights issues revealed in the summer of 2012 when labor unrest, child labor, several suicides and slave-like working and living conditions became public.

Whether they were directly or indirectly responsible for these deplorable labor violations and conditions, Apple's brand and reputation suffered a blow leading them to examine and revise their third party labor and human rights risk management.

This example demonstrates how reputation risk acted as an amplifier of another serious risk, in this case, labor and human rights risk.

Reputation risk and safety risk:
Sewol Ferry sinking in South Korea

A large industrial conglomerate that owns a passenger shipping company knows that providing for the safety of its critical stakeholders – employees and passengers – is a non-negotiable component of risk management. And so should the government agencies charged with monitoring and enforcing safety laws. Yet, the Yoo Byung-eun family, who owned such a company in South Korea, apparently did not know this nor did the government agencies charged with enforcing safety laws. The family and its companies are currently under investigation for the Sewol Ferry sinking, which apparently occurred due to massive safety violations from cargo overloads, leading to the death of 304 passengers (most of them of them teenagers on a field trip). The family patriarch disappeared right after the disaster, was the subject of a manhunt and was eventually found dead in one of his mansions. Government officials have resigned or been fired and the South Korean president has issued national apologies.

This case illustrates the massive amplifier effect that reputation risk has on another risk: safety risk. The unwillingness of the family, the company and the government to prepare properly for this risk will have long-term reputational consequences from broken stakeholder trust.

Reputation risk and cyber-risk:
The targeting of Target

Any company that handles thousands if not millions of customer privacy protected data – whether in banking, retail or healthcare – surely knows that they have a possible data breach/cyber-risk. If that company is found to have been lax, ignorant, negligent or otherwise non-compliant about safeguards, it is also predictable that such a company's reputation may be adversely hit.

That is exactly what happened to the US retail giant Target in the fall of 2013, when a fairly simple scheme allowed cybercriminals to collect 40 million credit card numbers and 70 million addresses, phone numbers, and other personal information of Target customers.

The revelations that followed severely hurt Target's reputation and stock

DISSECTING 'REPUTATION RISK'

price as well. This is again an example of how reputation risk can attach itself to another material form of risk – in this case cyber-crime risk – and potentially amplify its impact.

Other special qualities of reputation risk

In addition to the overarching fact that reputation risk attaches itself to other forms of risk, it is also special for other reasons. Let's review a few of these.

It's fast

> I'm going to Africa. Hope I don't get AIDS. Just kidding. I'm white.
> @JUSTINESACCO, 20 DECEMBER 2013.[22]

The now former Vice President, Public Relations, at IAC, a media company headquartered in New York City, who tweeted this just before boarding a flight from New York City to Johannesburg, suffered the consequences of reputation risk at the speed of light. She was terminated shortly after disembarking in Johannesburg.

Her company, in turn, demonstrated a deft handling of an increasingly all too familiar social-media based crisis when an employee communicates embarrassing, disparaging or damaging statements or behaviors on social media and a rapid and thoughtful response is necessary.

It can be positive

The musical group Coldplay took a calculated reputational risk with their early and deliberate release of three songs from their 2014 album *Ghost Stories* three weeks before the sale of the album, in exchange for payment by fans of the full price of the album.

Coldplay took the negative risk (to their reputation) that if the album was disappointing, they would lose fans for the next time. They also took the positive risk (to their reputation) that if the album was as good or better than past albums, they could do as well or even better than before.

The reputation risk actors (Coldplay in this case) and the reactions of their stakeholders (the buying public) determined whether the reputation risk they took was positive or negative. Judging from industry reports after the album was released, their gamble paid off.

It's contagious

The reputational impact of an event can reach well beyond its inner circle. The reach of the Snowden affair hasn't stopped with its reputational impact on the US government. By association, the US technology sector (including companies not involved in national security related technologies) has been very concerned about the reputational fallout on their competitiveness in the global marketplace, especially companies offering services in the cloud.[23]

It's tangible

The reputational impact of a negative event can be very tangible. Here's an example based on a US government annual report on global human trafficking:

> *Investors are dumping shares in some of Thailand's largest food companies as the shrimp-fishing titan faces the possibility of U.S. sanctions following a furor over the use of forced and child labor in its fishing supply chains.*[24]

Dr Kossovsky has written two excellent books on reputation and the

DISSECTING 'REPUTATION RISK'

impact of reputation risk on financial value, using extensive analysis and algorithms to measure the impact on stock value. He has concluded: 'reputational crisis will shave an average 7% from a company's market capitalization' and 'reputational value restoration can yield a net additional average of 13.5% in market capitalization'.[25]

It's cross-disciplinary
Reputation risk management is not the domain of one particular function, despite its traditional location within an organization's public relations, communications or public affairs office. While these functions continue to be critical to reputation management, the definition and examples of 'reputation risk' provided earlier should offer ample evidence that reputation risk is cross-disciplinary and requires a multi-functional approach. Figure 2 offers a general illustration of this concept.

FIGURE 2. Reputation risk is cross-disciplinary.

	REPUTATION RISK #1: CYBERCRIME
	REPUTATION RISK #2: HUMAN RIGHTS
	REPUTATION RISK #3: SUPPLY CHAIN
	REPUTATION RISK #4: CORRUPTION

Disciplines: ENVIRONMENT, HEALTH & SAFETY | CSR & CORPORATE RESPONSIBILITY | COMMUNICATIONS & PUBLIC AFFAIRS | AUDITING | FINANCE | GOVERNANCE | LEGAL | HUMAN RESOURCES | RISK MANAGEMENT | INFORMATION TECHNOLOGY | SECURITY & CRISIS MANAGEMENT

It's related to crisis management

Reputation risk like any other risk has an inextricable relationship with crisis management. A company without a proper crisis management plan is a company without a proper risk management program – they are two sides of the same coin. That can be said even more for reputation risk management because of its strategic, material and potentially volatile nature. The crisis management team and plan should be educated in and versed on reputation risk issues and include reputational concerns in all preparations for a crisis.

It's related to resilience and integrity

As we will explore in Chapter 6, reputation risk and the integrity of leadership and the organization itself are also intricately inter-related. It's simple: the more an organization has internal programs to boost its ability to deal with problems, risks and questions at their early stages, the more resilient, nimble and effective that organization will be in handling risk. A leadership that supports problem resolution is a higher integrity leadership by definition as they won't be afraid to hear the bad news and will encourage their employees to air issues early and often without fear of retaliation.

It's strategic

Reputation risk has recently and increasingly been acknowledged to be a strategic risk by boards, C-suite executives and CEOs.[26] This extends to any kind of entity. Just like a company, a university can ill afford to have a major scandal like Penn State University's child sexual abuse scandal of a few years ago which revealed troubling, systemic and high level flaws within the leadership of the university. Reputation risk can

DISSECTING 'REPUTATION RISK'

be sudden, swift and material and as such needs to be handled at the highest levels of an organization including the board regardless of form or mission.

See Figure 3 for an illustration of where reputation risk visually resides within the universe of risks.

FIGURE 3. Reputation risk is a strategic risk.

- ALL RISKS
- *Not all risks are strategic*
- *Strategic risk cuts across all risks*
- STRATEGIC RISK
- *Reputation risk cuts across all risks*
- REPUTATION RISK
- *Reputation risk is a strategic risk*

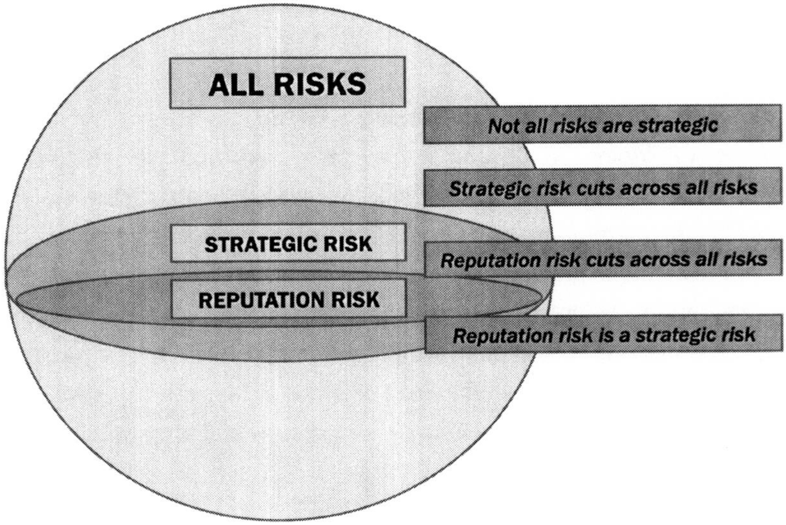

PART II
TRIANGULATING REPUTATION RISK

CHAPTER 3

Reputation Risk Within the Risk Universe

RISK MANAGEMENT IN GENERAL – how effective it ultimately is – depends largely on the organization, its sophistication, age, experience, geographical footprint, industry, sector, leadership, etc. In this chapter reputation risk management is placed within the risk 'universe' defined as risk architecture (the infrastructure of risk management) and the 'big buckets' of risk (the substance of risk).

Reputation risk and risk architecture

Figure 4 shows five stages of risk management evolution. It is only in the latter stages of development (namely 3–5) that effective reputation risk management is even possible. Let's quickly review these forms of risk management and understand how reputation risk management fits (or doesn't) within this evolutionary snapshot.

FIGURE 4. Risk management architecture: An evolutionary view.

REPUTATION RISK MANAGEMENT LIKELY TO OCCUR ONLY IN LATER STAGES

Stage 1: No risk management

The complete absence of any formal or even informal risk management program happens more often than one would think, certainly in the start-up phase of a for-profit company, for example. Frequently non-profits will not even think of risk in connection with what they do unless they have some regulatory or governmental relationship. Even more surprising, are the more evolved entities such as government agencies, academic institutions and even more mature companies that have not created formal risk management programs. Reputation risk management is an unknown concept to these entities until something bad happens to them.

Stage 2: Rudimentary or tactical risk management

This is the lowest common denominator of risk management. It applies mainly to entities that are truly small or in the very earliest stages of development. Rudimentary or tactical risk management consists of a haphazard collection of policies or approaches to risk where only known risks are acknowledged and discussed sporadically by management and/or the board, but are not subject to a systematic or formal program. There is no attempt to pull it all together or to have a related crisis management plan. As in Stage 1, reputation risk management is an unknown concept until something bad happens.

Stage 3: Basic ERM

At this stage, an entity has recognized the need for a cohesive approach to risk management. This could include a sophisticated start up, NGO, mid-size private or public entity. This entity has probably experienced the adverse impact of risk-related incidents or crises and is interested in learning lessons and creating appropriate systems to identify and mitigate risks.

Both management and the board are paying attention to risk management but are not likely to have deep expertise, save in the case of one or a couple of members of senior management as well as the board. There is no dedicated risk committee at the board level yet but the audit committee is actively looking at risk oversight as part of its oversight portfolio.

There is likely to be a senior executive in charge of risk management – such as a chief risk officer, for example – though at this stage it is more likely to be a many-hatted executive like a general counsel, chief financial officer, audit officer, compliance and ethics officer.

It's possible that reputation risk management has been identified as part of the portfolio of risks but unlikely that a sophisticated approach to reputation risk management is in place yet.

Stage 4: Developed ERM

This is a more evolved entity with a certain threshold number of operations and/or revenues (bordering on $1 billion or more), with a diverse and extensive footprint (multiple countries, multiple businesses); it can be public or private.

This entity has a definite understanding of its risk universe, has created a sophisticated framework to handle risk in a proactive and periodic manner. It has most definitely experienced risk-related incidents or crises and has applied lessons learned to mitigation and prevention.

There is likely to be a chief risk officer and a risk team, whether stand-alone or within an organizational matrix. Systems and experts are actively engaged in providing definition, framework, expertise and focus to risk management. It's likely that reputation risk management has been identified as part of the portfolio of ERM risks that are looked at and dealt with systematically.

Both management and the board are paying attention to risk management, have risk issues regularly on their executive and board agendas and require regular ERM updates and discussion in the boardroom. The board itself is likely to not only look at risks from the audit committee's perspective but also from other committee standpoints. In fact, the board may have created a separate risk committee either because of regulatory exigencies or because of the complexity of the business and risk environment.

Stage 5: Strategic risk management

Building on everything stated about Stage 4, strategic risk management is the most evolved form of risk management where both management and the board have a sophisticated and well informed view and infrastructure in place to deal with all entity risks including reputation risk which is fully incorporated not only into risk management but into the entity's long-term strategic and annual business planning.

Reputation risk in the pantheon of 'big buckets' of risk

Now that we have reviewed the various stages of evolution of risk architecture or infrastructure, let's turn to a view of the major categories or what I like to call 'big buckets' of risk, substantive areas that most entities must deal with in one shape or another. Of course, not all categories apply to all entities and there are others not included here that are applicable to certain sectors or industries. The point is not to be comprehensive but to illustrate where reputation risk ultimately fits into this pantheon of 'big buckets' of risk.

Reputation risk is not an additional 'big-bucket' category but a different kind of crosscutting category – an altogether different animal that is nevertheless interconnected with the other eight (or more or less) categories.

Political risk

This category includes risks emanating from the particular location, its government (whether local, provincial, state, national, federal or multi-national), and the state of political and socio-political affairs in that location. It includes the risk of political interference with contractual and international laws, norms and regulations, and the risk of breakdown in the rule of law or of an unpredictable or corrupt judicial system. Political risk also includes risk of war, conflict and civil unrest.

Operational risk

This category includes risks embedded in operations – the overall administration and organization of the entity, whether it is highly decentralized or centralized, its workforce, labor and payroll management, supply chain and procurement program, business continuity and crisis management, distribution, sales and other business development channels, joint venture and partnerships, management of local operations, etc.

**REPUTATION RISK
WITHIN THE RISK UNIVERSE**

Financial risk

This category includes risks relating to financial management and the internal plumbing of an organization's financial structure, reporting, debt and leverage, financial/contractual inter-relationships, internal controls, taxation, sales and business development, financial incentives, currency exchange, hedging and derivatives, state of the national and international economy, the precise details of how all transactions and financial details are collected, reported and disclosed along the financial reporting chain, etc.

Legal, regulatory & compliance risk

This category includes risks emanating from anything and everything that can be categorized as a government sanctioned law, regulation or compliance requirement in any jurisdiction anywhere. From corruption to fraud, from privacy to data security, from discrimination to harassment, from anti-trust to anti-money laundering, these laws exist almost everywhere, in different shapes and forms, at different levels, different regulatory and enforcement regimes and within different jurisdictions and venues.

Environmental, health, safety & quality risk

This category encompasses all things environmental, health, safety or quality, which which can differ greatly depending on the industry or sector involved.

Supply chain risk

This is a relatively new, complex and high impact multifaceted set of risks beginning at the inception of a supply chain (including risks relating to a possible multitude of third parties), including a variety of other risks along the way (slave labor, child labor, human rights risks) and culminating with products or services with additional risks (integrity, quality, safety of product or service).

Technology risk

This category includes all information security, cyber, technology, software and hardware related risks, which seem to be multiplying and expanding on a daily basis.

Integrity and culture risk

This category is the risk of leadership, people or culture failure. A set of risks few if any ERM practitioners think about or systems incorporate. A critically important component of this bucket of risk is the risk of leadership malfeasance and culture failure which often go together and which can have devastating reputational consequences on a company, its people, products, services and other stakeholders. The risk of leadership and culture failure are arguably the two biggest risks companies never analyze or prepare for in advance.[27]

Reputation risk is different

Why doesn't reputation risk merit its own special 'big bucket' category? Because it doesn't happen in a vacuum – it happens in conjunction with one or more other risks. Figure 5 illustrates how reputation risk is not a stand-alone risk but one that cuts across all other substantive 'buckets' of risk.[28]

REPUTATION RISK WITHIN THE RISK UNIVERSE

FIGURE 5. The 'big buckets' of developed ERM & reputation risk.

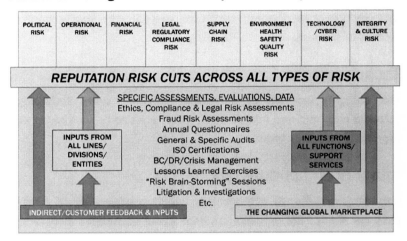

CHAPTER 4

Reputation Risk Actors & Stakeholders

Reputation risk and you

REPUTATION RISK HAPPENS at the most elemental level: you. Ask yourself the question: if something negative or positive occurs to me (either by happenstance or because I have done it to myself) what are the possible consequences? The alternatives are pretty simple: my reputation may suffer or improve.

Put a little differently: who are my stakeholders? Who and what can affect me and who and what can I affect from a reputational standpoint?

Figure 6 provides a snapshot of the many possibilities that may exist in an individual's world to be affected by and /or to affect the reputation of others.

A few illustrations might help:

- You live in a neighborhood that is known for its wealthy homes and families – whether you are wealthy or not, people will think you are wealthy. Your neighborhood (and therefore you) has a reputation for being wealthy.

- You live in a city that is known for its high crime statistics. Whether

REPUTATION RISK ACTORS & STAKEHOLDERS

FIGURE 6. Reputation risk and you.

Nodes connected to "You":
- Your Region or Geographical Area
- Your Sports Team
- Your Political Affiliation
- Your Political Candidate
- Your Friends
- Your Clubs & Associations
- Your Neighbors
- Your Country
- Your Ethnicity/Nationality
- Your School
- Your Family
- Your President or Prime Minister
- Your Company/Employer
- Your Job
- Your Community
- Your City
- Your Government

your particular neighborhood or not experiences high crime, people from other locations may not visit your city or may think that your entire city has high crime. Your city (and therefore your neighborhood indirectly) has a reputation for having high crime.

- A member of your family commits a crime and is incarcerated. Though you had nothing to do with this criminal activity, your reputation suffers by association. You and your family's reputation are tainted by your relationship to the convicted criminal family member.

- Your company has been fined repeatedly for violating the law.

Whether you have taken part in these violations or not, as an employee of this company you may be tainted by association. Your company (and maybe you) has a reputation for non-compliance or worse.

Reputation risk and the organization: actors & stakeholders

Now let's tackle reputation risk management at the entity or organizational level. In any organization, there are reputation actors and reputation stakeholders. Figure 7 shows the variety of reputation stakeholders that an entity might have.

Reputation actors

We are all reputation actors. Within the context of an organization, who is a material reputation actor? Material reputation actors are those who have a range of abilities and powers to affect the reputation of their entity for better or for worse. Thus, not all employees will necessarily have a material reputational impact on their employer but they can if they take the right or wrong steps. However, there are certain actors with greater reputational impact powers than others. These would include:

- **Employees and managers** who are on the front lines of a business with customers, partners, and interacting with key stakeholders like regulators and the media, for example. You can't be too small to have a material impact: the sandwich maker at Domino's Pizza who was videotaped doing nasty things to a sandwich before selling it to a customer had a material impact on Domino Pizza's reputation for sanitary conditions.

REPUTATION RISK
ACTORS & STAKEHOLDERS

- ***Senior executives and the C-suite*** who are both on the front lines and strategic. The annals of scandal history are littered with examples of leadership stupidity, malfeasance, misbehavior and downright criminal activity affecting the reputation of an entity.

- ***Board members*** who are not necessarily on the front lines (though they could be) but have a key strategic role in oversight. While there may not be many examples of malfeasance of an entire board, there are a few of board members behaving badly and affecting the reputation of the entity.

The common thread that runs through each reputation risk actor is that each one, to a greater or lesser degree, has the ability and power to affect the reputation of the entity – for good or bad:

- The employee who makes a disparaging tweet may damage his/her company's reputation.

- The employee who does a great job in leading an amazing community program for his or her company will surely enhance the company's reputation with that community.

- The executive leadership of a large public university (Penn State) who looks the other way and proactively hides the criminal misbehavior of a star sports coach for over a decade severely damages the university's reputation when the facts become public.

- A CEO (Starbucks) who decides to change his company's wage policy to elevate the quality of life and esprit de corps of employees elevates the company's reputation for fair treatment, trust, loyalty and respect with employees and other key stakeholders like customers.

- Board directors who improperly oversee the safety record of a mining company (Massey Mining) contribute to the serious reputational impact of a deadly mining accident on the company.

- Directors who, after much deliberation and investigation, agree on terminating a CEO found to have committed serious misbehaviors (American Apparel) will help an entity protect and/or reinstate its former better reputation.

- The board's firing of the CEO after the revelation of serious conflicts of interest can help send the stock to new heights after the departure is announced (Chesapeake Bay Energy, stock up 9%).

Reputation stakeholders

There are internal and external stakeholders – people or entities with a stake in what another entity or person is doing for or with them. Some stakeholders are both internal and external and some stakeholders may also be actors (e.g. boards of directors). The bottom line is that every entity has multiple stakeholders all of whom are dependent upon the entity's reputation and capable of influencing the entity's reputation (see Table 5).

REPUTATION RISK
ACTORS & STAKEHOLDERS

TABLE 5. A multitude of organizational reputation stakeholders

Internal	External

• Owners/shareholders/investors:
  - Family
  - Private
  - Public
  - Government
  - Institutional
  - Activist hedge funds
• Boards of directors, trustees or supervisors
• Board committees & chairs
• Council of advisors
• Employees
• Temporary & contract workers
• Labor unions
• Workers councils
	• Customers, purchasers and clients
	• Users of products and services
	• Prospective owners, shareholders, investors
	• Partners & suppliers
	• Communities
	• Non-governmental organizations
	• Prospective employees
	• Government agencies, regulators, enforcers:
	  - Local
	  - Provincial/state
	  - National
	  - International
	• Media & social media

An entity's critical actors create expectations in the entity's stakeholders. A crucial role of stakeholders in reputation risk is that they hold the keys to the entity's reputational trajectory. The critical link between stakeholders and the entity is trust – trust in its promises, future behaviors, prospects, performance and results. When that trust is broken (through a reputational hit, for example), that stakeholder may withdraw support or stop interacting with the entity. If it's a partner, the partner may dissolve the relationship; if it's a government, the government may start an investigation; if it's an investor, the investor may cash out of the entity.

Stakeholders depend on the delivery of those expectations – this is what trust is all about. If a negative event occurs (downgrade in financial prospects, opening of material government/regulatory investigation, misbehavior by the CEO), it can affect stakeholder trust negatively and therefore possibly affect the financial well-being of the entity as well.

On the other hand, if expectations are managed rationally and something better than expected happens in terms of performance, results, behaviors, the entity might get a reputation and concomitant financial boost as stakeholder trust increases.

The reputation risk 'Rashomon' effect

To add to the confusion about the already amorphous and malleable nature of reputation risk, reputation itself is a curious thing – it is in many ways in the eye of the beholder. Thus within and outside of an organization, what reputation risk means, whom it means what to, can be multifaceted and diverse. The core of an issue or problem an organization confronts may be subject to multiple reactions, interpretations and gradations. This is a critical reason why companies and other entities should have an informed and organized approach to managing potential reputation risk.

The classic movie *Rashomon* by the great Japanese director Akira

REPUTATION RISK
ACTORS & STAKEHOLDERS

Kurosawa offers an interesting and useful lesson. Five people who are witness to or participants in an event tell five different and sometimes colliding stories of what happened. At the end, it is unclear what the 'real' or 'true' set of facts actually is as each person (each stakeholder in the story) remembers it differently either because they want to see it in a certain way or because their memory has failed them. Or maybe it is because as imperfect humans, we simply do not and cannot remember everything about something (certainly not in the same way as our fellows) and thus the facts and the reputation implications of those facts are remembered, interpreted and reported differently.

Something similar happens in the corporate or organizational world – each of us, as limited humans, only have so much capacity or ability to interpret and understand what happens and we bring our own lens to the events. Plus, we usually only know one sliver of the facts – good, bad, or indifferent; few of us know everything about something. Thus when a set of facts occurs to an organization, stakeholders react with their own lens and perspective, knowledge and limitations. Thus there is a risk – negative or positive – that multiple stakeholders will have multiple interpretations of the facts that will have an impact on the reputation of an organization.

Take the BP Deepwater Horizon environmental disaster. Once it occurred, a vast number of stakeholders were affected by it in one way or another and had a variety of different interpretations and reactions to the situation as the many lawsuits, investigations and unrelenting media coverage have revealed. Among the stakeholders and their not always aligned reactions were the following:

TABLE 6. Reputation risk 'Rashomon' effect example: Stakeholder interpretations & reactions to the BP Deepwater Horizon disaster

Stakeholder	Range of reactions/interpretations
Employees & their families	Anywhere from deep anger and distrust (families of dead and injured) to defending the status quo (BP did its best and these things can happen)
Partners	Each of the partners to the project (BP, Transocean, Halliburton) blamed the other partner(s) to varying and differing degrees
Community	From legitimate grievances for damage restitution to misrepresentation, fraud and theft in order to collect undeserved restitution money
Protected habitats	Claims of varying levels of destruction of wildlife and protected natural habitat
Customers	Anywhere from no reaction to boycotting BP at the gas pump
Investors	Anywhere to staying invested in BP for the long term to divesting immediately
Regulators	From possible previous laxity in enforcing regulations to full-on law enforcement
Media	From accepting the statements of BP officials to deeply critical investigations

This is the reputation risk 'Rashomon' effect: as the facts of a reputational event unfold, different stakeholders may have differing interpretations and reactions which only add to the possible negative or positive reputational impact of that event.

REPUTATION RISK ACTORS & STAKEHOLDERS

The reputation risk 'Rashomon' effect is yet another core reason why reputation risk is and must be handled at the highest levels of an entity as a strategic risk and must be part of any preventative planning and real-time crisis management.

FIGURE 7. The reputation risk 'Rashomon' effect.

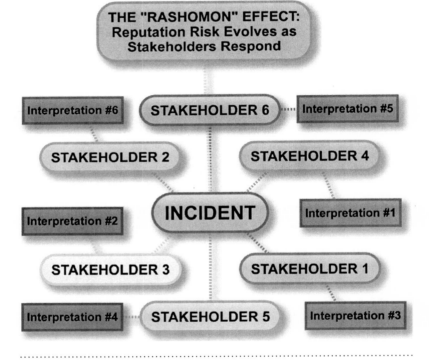

Reputation risk ownership and oversight

This discussion of actors and stakeholders and emphasis on the strategic nature of reputation risk leads to two important questions:

- Who owns reputation risk management?
- Who is responsible for reputation risk oversight?

One thing is clear about the state of reputation risk management in most organizations today: there are either no cooks or too many cooks in the kitchen and, often, there is no chef. This is not a good state of affairs given the strategic nature of this risk.

Reputation risk ownership: Designing the right framework

Depending on the state of development of its risk management infrastructure (as discussed in Chapter 3), an entity needs to find the best location for its reputation risk management. This in no way requires the construction of another function or committee but instead a creative approach to finding existing, preferably multi-functional, resources within the organization equipped to deal with these issues.

Starting from the top – the board and the CEO – an entity should consider some of the following ideas to assign reputation risk ownership and management within the entity:

- *Team* – One of the better locations to assign reputation risk management is an already established high-level, cross-functional team, like a corporate responsibility and/or risk management committee.

- *Ownership* – Because of the complexity and ubiquity of reputation risk, there shouldn't be a single reputation risk owner but maybe several internal experts – such as investor relations, ethics and compliance and public relations and communications executives who should share ownership of this issue.

- *Strategy* – Any strategic discussion, planning or development session or initiative should include a reputation risk management component.

- *Crisis* – The organization's crisis management plan and team should have a reputation risk management component – and an expert internal or external resource that has been identified for crisis situations. This could be public relations, investor relations or communications executive or someone else qualified.

- *Risk* – Companies with a well-established internal risk management program should consider assigning top executive management responsibility for reputation risk management. An idea whose time may have come for some highly evolved organizations is that of the Chief Integrity, Risk & Reputation Officer discussed below.

In a 2011 piece, Jorge Cachinero makes a strong case for the creation of a Chief Reputation Officer tied to the rising importance of reputation as an intangible asset. Because of this, reputation management cannot remain strictly the domain of public relations and communications but must graduate to a more holistic, executive and board-reporting role. I would go a couple of steps further to suggest that some companies should consider the creation of an even more inclusive role – a chief integrity, risk and reputation officer (CIRRO) who would act as the executive in charge of a holistic strategic approach to these issues, their coordination and relationship to entity strategy, reporting to both the

CEO and board. The CIRRO would serve as the chair of a global integrity, risk and reputation committee where all manner of related issues would be handled and in-house experts would periodically coordinate policy, tactics and strategy on these issues. See Figure 8 for an illustration of such a structure.[29]

FIGURE 8. Reputation risk ownership framework: A future model?

Reputation risk oversight: The responsibility of the board

Boards have a critical role to play with regard to reputation risk. They own risk oversight and, as a strategic risk, they therefore also own reputation risk oversight. If a board doesn't have reputation risk on its agenda, it is being irresponsible because it is missing a critical component of responsible risk oversight.

Boards hold the CEO and executive team accountable on a broad variety of strategic issues. They have committees specialized in different aspects of management of the entity to do this. Depending on the entity, audit,

REPUTATION RISK
ACTORS & STAKEHOLDERS

risk, and/or compliance committees own risk oversight. If they haven't already, as part of their oversight portfolio, such committees need to add reputation risk oversight to their portfolio. Boards would also benefit from having at least one independent director, savvy and experienced in risk issues, not just tangentially but directly.

The board is the ultimate protector and guardian of organizational integrity and value. Reputation loss (and gain) can materially affect integrity and value. What's more, board members have their own personal integrity and value to protect. Ultimately, reputation risk oversight is one of *the* intrinsic governance roles of a board. The new bottom line for boards is that they need to demand a new bottom line from their chief executives – one that provides as much accountability on risk and reputation management and metrics as is currently provided on financial management and metrics.

To gain the upper hand on reputation risk oversight, boards should consider the following four maxims:

1. *Accountability* – Require that management tackle reputation risk head-on at both the C-suite level and enterprise risk management levels and that metrics to measure both financial and non-financial (ESG) issues are deployed.

2. *Inclusiveness* – Insist that reputation risk be one of the core strategic risks within the entity's risk pantheon and not just a post-crisis afterthought.

3. *Cross-functionality* – Ensure that the entity undertakes a cross-functional approach to reputation risk management, taking advantage of all internal expertise, making sure it isn't simply

placed into one (or no) functions or silos, and making use of necessary outside expertise.

4. *Preparation* – Management and the board should be kept up to date on the latest reputation risk trends, tools and developments, including undertaking reputation risk crisis workshops. The board should ensure that this is happening by asking management the right questions and learning from experts what those questions might be.[30]

....

PART III
DEPLOYING REPUTATION RISK

CHAPTER 5

Reputation Risk Strategies and Toolkit

WE FIRST EXPLORE reputation risk management strategies potentially available to an entity and then look into fifteen tactical tools for effective reputation risk management. Then we determine who, amongst an entity's reputation actors, is most likely to be responsible for and/or deploy the tools in the toolkit.

Reputation risk management strategies: a typology

Now we turn to a typology of possible reputation risk management strategies an entity may have at its disposition. It is based on two factors: 1) what, if anything, does the entity have in place in terms of risk management architecture in general (how evolved is its risk management system as reviewed in Chapter 3), and 2) how deep and extensive is the knowledge of, and capability to, manage reputational risk issues within the entity. See Figure 9 for a visualization of this analysis.

This analysis yields four possible types of reputation risk management strategy: the 'incipient', the 'chaotic', the 'façade' and the 'effective'. Let's take a look at each of these types and then consider the various tactical tools or elements that should be part of an effective reputation risk management strategy.

FIGURE 9. Reputation risk management strategies: A typology.

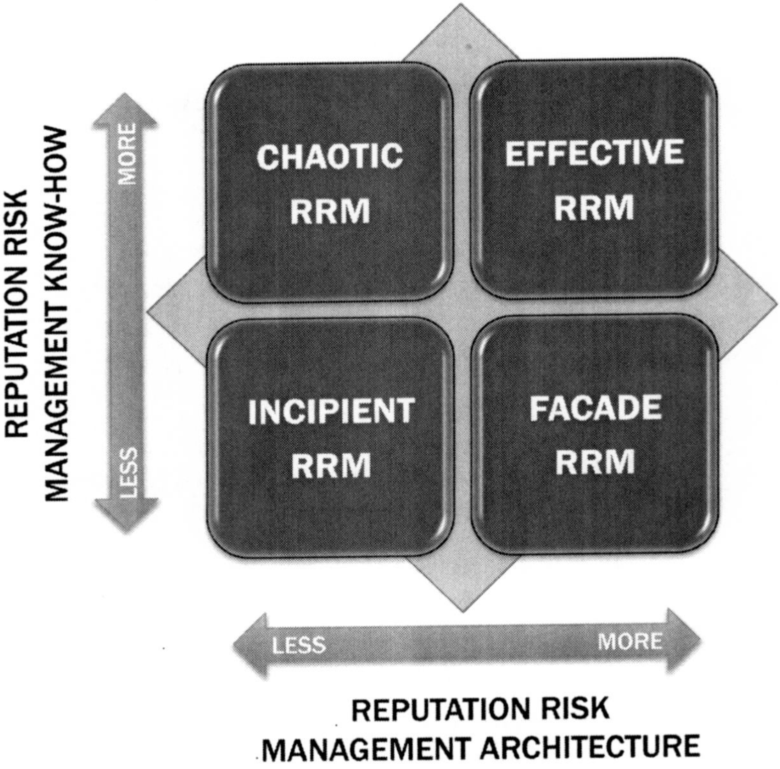

Vertical axis: REPUTATION RISK MANAGEMENT KNOW-HOW (LESS ↔ MORE)
Horizontal axis: REPUTATION RISK MANAGEMENT ARCHITECTURE (LESS ↔ MORE)

- CHAOTIC RRM (more know-how, less architecture)
- EFFECTIVE RRM (more know-how, more architecture)
- INCIPIENT RRM (less know-how, less architecture)
- FACADE RRM (less know-how, more architecture)

The 'incipient' reputation risk management strategy

An organization does not yet have a risk management program or deep know-how in place. This could be because the organization is relatively new (a start-up), relatively small (a family business), relatively local (a grocery store), relatively unregulated (in the professional services field) or relatively high risk (driven by a 'cowboy' business culture). Whatever the reason, the organization does not believe in or does not know it needs a formal risk management program, let alone a reputation risk management strategy.

The 'chaotic' reputation risk management strategy

This is an organization that understands, to a certain extent, that it faces certain risks, including reputation risk, but does not have a formal program, policy or internal architecture in place to deal with such risks. It may draw upon the expertise of one or more internal experts when needed. So, for example, if a manufacturing site has an environmental incident, it may or may not call on its environmental manager to help solve the crisis. This is a chaotic and disorganized approach to risk management that only touches upon the issue of reputation risk by chance.

The 'façade' reputation risk management strategy

This approach is more about form and less about substance. An example might include a global company that undertakes an enterprise risk management exercise every year, asking its local managers to fill out questionnaires or surveys, but not effectively applying expert analysis or deriving action plans to understand and resolve the risks that exist. It is in effect a bureaucratic information gathering exercise, lacking in follow up or solution execution. With façade RRM, it is more likely that a PR, communications, sales or marketing person is involved with reputational issues without the deeper and broader expertise needed for effective reputation risk management.

REPUTATION RISK STRATEGIES AND TOOLKIT

The 'effective' reputation risk management strategy

It is only when we reach the combination of deep risk know-how within an organization, including how to tackle reputation risk, and the existence of a framework and infrastructure to enable the optimal use of that know-how through an evolved ERM system that we reach what can be called effective RRM.

Let us now turn to the reputation risk management toolkit – fifteen tactical elements, most of which are present in an 'effective' reputation risk management strategy, to understand more fully what it means to reach that point.

The reputation risk management toolkit

The deployment of reputation risk management will be as good as an organization's approach is to risk management generally and to strategic risk management specifically. Because of its unusual nature (as described in Chapters 1 and 2), reputation risk is different from other risks and should be handled accordingly – as a strategic risk and integrated into strategic risk and ERM programs, if they exist. Even an organization that is just beginning to build its risk management program would be wise to place reputation risk at the very top of its strategic risks for the many reasons elucidated throughout this Handbook.

A well-rounded reputation risk management strategy (the 'effective' one described above) will contain most of the elements listed below; however, these elements should not be considered a new or stand-alone program. These elements should be integrated into the fabric of the organization within systems and procedures that already exist. Much like reputation risk, a reputation risk management program should be layered on and complementary to existing processes and programs. In fact, the toolkit can also be viewed as a reputation risk checklist against which to compare how an entity is or is not integrating reputation risk considerations.

That said, the following elements are part of a well developed, effective reputation risk management strategy:

1. **Risk program**

    No matter its size, nature, mission or strategy, an entity should have some type of formal risk management program in place.

    The key question management and the board should be asking is: what is appropriate for our entity's stage of development, footprint and strategy? For some ideas, see the 'Risk architecture' discussion in Chapter 3.

2. **Assessment & monitoring**

    Critical to the success of any risk management program is assessing the top risks, monitoring them, developing policies and approaches to solving them.

    With respect to reputation risk assessment and monitoring, among the key questions management and the board should be asking are:

    - Does the organization know what its possible reputation risks are?

    - Has a reputation risk assessment been conducted as part of an overall risk assessment asking specific reputation risk vulnerability questions? For example:

**REPUTATION RISK
STRATEGIES AND TOOLKIT**

- Does the entity have an understanding of possible or likely reputation risk issues it could face, based on its past history or industry benchmarking?
- Has senior management conducted crisis management exercises that include realistic and relevant reputation risk components?

3. **Knowledgeable & appropriate resources**

   Internal and external resources appropriate to the size, nature, mission and strategy of the entity need to be in place.

   Among the key questions management and the board should be asking are:

   - Is someone – a high level executive – providing overall executive guidance on reputation risk and minding the reputation risk store?
   - Are one or more internal experts paying attention to the nuances of reputation risk as it relates to the organization?
   - Have these resources been properly trained and equipped to spot, understand and manage reputation risk?
   - Is there an adequate budget to manage reputation risk?

4. **Cross-functional approach**

   Given its unique qualities (see Chapter 2), a critical component to successful reputation risk management is a cross-functional approach.

   Among the key questions management and the board should be asking are:

- Do we have a high-level cross-functional team already in place that is tackling or can tackle reputation risk issues?
- Do we have the right internal reputation risk experts and are they part of this cross-functional approach?
- Do they have access to appropriate resources including external experts?
- Do they have access to the right high levels of the organization?

5. **Front line integration**

    A critical component of reputation risk vigilance is to reach business teams on the front lines with useful and practical guidelines and examples of reputation risks and opportunities.

    Among the key questions management and the board should be asking are:

    - Are supervisors properly equipped to identify and deal with reputation risk issues at the front lines of the organization?
    - Do supervisors and managers know what to do when a reputation risk hits? Do they know where to turn for help?
    - Do they know how to identify reputation risk opportunities?

6. **PR & communications integration**

    Reputational issues can come up on a regular basis especially with the advent of social media crises and mini-crises, which can blow up into larger crises if unattended, requiring specialized training and vigilance within an entity's PR and communications team.

REPUTATION RISK STRATEGIES AND TOOLKIT

Among the key questions management and the board should be asking are:

- Is there a communications plan in place for reputation related communications – internally and externally?

- Has the PR and communications team been properly trained on reputation risk issues?

- Has a team been identified to tackle the smaller, reputation mini-crises? Do they know when to escalate to the full-fledged crisis management plan and team?

7. **Clear policies & guidelines**

Reputation risk must be properly addressed in relevant documents, policies and procedures including mentions in a code of conduct and helpline/hotline instructions.

Among the key questions management and the board should be asking are:

- Is the approach to reputation risk management documented clearly and accessibly to the necessary internal actors and stakeholders within the organization in all the relevant documents?

- Are the guidelines written in clear and actionable language?

8. **Education & training**

Management and the board should become savvy about the potential reputational crises that can befall their organization through periodic risk training.

THE REPUTATION RISK HANDBOOK:
SURVIVING AND THRIVING IN THE AGE OF HYPER-TRANSPARENCY

Among the key questions management and the board should be asking is:

- Are we getting the best advice and education from both internal and external sources to maintain proper management and oversight of reputation risk with regard to our organization?
- Are we learning from our own reputational crises and incorporating these customized cases into our internal educational platform?

9. **Problem resolution/speak-up culture**

   Whether it is through existing helpline or hotline channels or through a separate employee or third party method, reputation risk management is greatly served by a culture that encourages early problem detection and speak-up without retaliation.

   Among the key questions management and the board should be asking are:

   - Do we have the appropriate safe to speak up/problem resolution mechanisms in place and are they well publicized?
   - Do we have the right culture of transparency and non-retaliation where mistakes are forgiven and problem solving is rewarded?

10. **Crisis management plan integration**

    No reputation risk management is complete without full integration with an entity's crisis management plan and team.

    Among the key questions management and the board should be asking are:

REPUTATION RISK
STRATEGIES AND TOOLKIT

- Is there an effective crisis team and plan in place?
- Is proper crisis management training being periodically conducted including consideration of reputation risk events?
- Are there members on the crisis management team equipped to handle reputational issues and/or have outside experts been identified to assist?

11. **Post-event SWOT (strengths, weaknesses, opportunities, threats) analysis**

   No good crisis should be wasted, so smart entities learn from their reputational hits or mistakes and build lessons learned into their future processes.

   Among the key questions management and the board should be asking are:

   - Once a reputation hit or crisis has taken place, is there a lessons learned exercise or SWOT analytical framework available to sort out what happened and provide improvement recommendations going forward?

12. **Issue rapid deployment force**

   Once the reputational crisis has hit or a reputational risk has been identified, an entity should have an immediate plan and team ready to handle and solve the root cause issue related to the reputational hit.

   Among the key questions management and the board should be asking are:

- Has a 'rapid deployment force' approach to issue-resolution been created and do we have the necessary internal and external resources to achieve efficient issue solution?
- Is the right team in place for the specific root cause issue identified?

13. **Strategic & business plan integration**

    Reputation risk is strategic. Executive teams should incorporate reputation risk and opportunity considerations into their strategic planning initiatives and annual business planning exercises.

    Among the key questions management and the board should be asking are:

    - Have we elevated reputation risk to a strategic risk in our annual and long-term strategic planning?
    - Have we integrated reputation risk into our board risk oversight agenda?

14. **Board oversight**

    Because reputation risk is strategic it should be a key consideration of board risk oversight, whether considered within an audit, risk, compliance or other committee.

    Among the key questions the board and board committee should be asking is:

    - Do the CEO and the executive team go beyond simple lip service to reputation risk and actually integrate it into strategic risk management, annual planning and long-term strategic planning?

15. Industry benchmarking and action

It's very important to know what others are doing for reputation risk management – both in general and within your sector. Additionally, an entity may also want to take concerted action with other entities on common reputational concerns. Witness the two reactive initiatives the global retail industry undertook after Rana Plaza and the more proactive approach of the auto industry in its adoption of the Supply Chain Ethics Declaration.[31]

Among the key questions management and the board should be asking are:

- What are other entities in general or within our sector doing to tackle reputation risk management?

- What initiatives exist within our sector or generally that we may want to join for reputation enhancement reasons (either because of a particular reputational hit or to stave off reputation risk contagion within the industry)?

- Why are we or why aren't we part of these initiatives?

- Should we consider starting an initiative that would not only address an existing reputational risk but also potentially provide a reputational enhancement and even competitive advantage?

Table 7 provides an overview of the responsibilities of the principal actors within an entity for each of these tools on a scale of leadership/material responsibility (■), substantial responsibility (■), awareness or knowledge responsibility (■), and little or no responsibility (■).

TABLE 7. Reputation risk management toolkit: Organizational actors & their primary tools

Element	All employees	Supervisors	Executives	CEO	Board
1. Risk program					
2. Assessment & monitoring					
3. Knowledgeable resources					
4. Cross-functional approach					
5. Front line integration					
6. PR & communications integration					
7. Clear policies & guidelines					
8. Education & training					
9. Problem-resolution/ speak-up culture					
10. Crisis management plan integration					
11. Post-event SWOT analysis					
12. Issue rapid deployment force					
13. Strategic & business plan integration					

REPUTATION RISK
STRATEGIES AND TOOLKIT

Element	All employees	Super-visors	Execu-tives	CEO	Board
14. Board oversight					
15. Benchmarking & industry action					

▧ = material/leadership responsibility; ■ = substantial responsibility;
▨ = awareness/knowledge responsibility; ░ = little or no responsibility.

CHAPTER 6

Optimizing Reputation Risk Management: the Next Strategic Imperative

THIS FINAL CHAPTER offers a framework for the selection of the best reputation risk management option for an organization. To do this, we take a slight detour into understanding the types of leadership and organizational culture that an entity can have when it comes to issues of integrity. Why? Because an organization's integrity culture and leadership style will heavily determine the type of reputation risk management strategy it will be able to deploy, if at all.

We then conclude with a big picture visualization of how reputation risk impacts an organization and a look into the future of reputation risk management where smart companies and other organizations are deploying reputation risk lessons learned to create new value.

How to optimize reputation risk management

As discussed earlier in this Handbook, reputation and integrity are two deeply intertwined matters. Two key ingredients relating to integrity go into understanding how to optimize reputation risk management for an organization: 1) what style of integrity leadership does the CEO exhibit?; and 2) what kind of integrity culture does the organization have? Let's take a look at each of these elements and then circle back to understanding the type of reputation risk management strategy that is possible.

First criterion: Integrity leadership

One thing is abundantly clear: sizeable scandals do not originate at the shop floor or the accounting department, they are not schemes concocted by low- or mid-level employees. Sizeable scandals begin and end at the top: with executives setting the wrong example or demanding impossible results; chief executives preaching 'do what I say, not what I do'; performance metrics pushing employees to achieve impossible and unsustainable financial targets; or CEOs holding dramatic carrots and sticks over top performers who, in turn, pressure and drive their staff to unrealistic and, in some cases, illegal results.[32]

A leader who is focused on pure bottom line results (whether these are financial or other metrics) without substantive regard for how those results are achieved, is a very different leader from one who pursues a holistic stakeholder approach where the financial bottom line is complemented by other considerations like ESG related metrics and balanced performance incentives.

When leadership experts, organizational behaviourists, ethicists and risk managers look at what creates (or destroys) a culture of integrity and accountability, a major focus, rightly, is on tone at the top. What leaders say and especially what they do has broad and pervasive effects on the organization and its people. This is especially the case with chief executives. What the chief executive says is closely listened to – what the chief executive does is even more closely emulated.

THE REPUTATION RISK HANDBOOK:
SURVIVING AND THRIVING IN THE AGE OF HYPER-TRANSPARENCY

Good governance is a key to getting this issue right. Boards need to get savvy to this behavioural dynamic, and, in turn, hold leaders accountable for creating and maintaining an effective culture that ultimately upholds and furthers the entity's reputation. There is a spectrum of integrity leadership styles that chief executives exhibit and which boards of directors should be attuned to:

The irresponsible leader – This CEO is at best oblivious and at worst hostile to issues of integrity and the importance of an ethical culture. Doing business with integrity is simply not part of his/her worldview. In fact, it's an oxymoron. The disgraced leaders of the companies involved in major scandals – Enron, WorldCom, Lehman Brothers – would fit into this category.

The 'irresponsible' leadership style is most likely to occur in companies that haven't yet felt the sting of a scandal – that haven't, yet, felt the blow to the entire organization, its stakeholders and bottom line.

The superficial leader – This is a chief executive who 'talks the talk' – promotes attractive marketing and branding communications with good words about ethics, integrity, responsibility, customer care, sustainability and stewardship. While this may create a pretty façade it doesn't support a deep culture of responsibility or integrity. In this culture, most employees (and other stakeholders) realize it's all about marketing and PR.

The superficial style of leadership does not institutionalize integrity – by providing a culture where it is safe for employees to speak up, for example. Instead, superficial leaders create a skin-deep 'Potemkin village' approach to integrity issues. Many of the leaders of the biggest global financial institutions involved in the recent financial scandals fit this category, as they often have well-developed, beautifully branded and expensive corporate responsibility and compliance programs, but rarely do they have a sustainable culture of integrity that reaches more deeply and broadly into the organizational machinery. Sadly, a large portion of business leaders probably fit into this category and little do they know that they may be leaving real, measurable value on the table by not moving up the evolutionary ladder to the next two categories of leader.

The responsible leader – These leaders actually 'walk the talk' and mean what they say about ethical culture and corporate responsibility. They put their money where their mouths are by developing leaders and employees, catering to major stakeholders such as customers, providing appropriate resources to promote and build integrity and corporate responsibility programs and incentives within the organization.

This leader has recognized the importance of having a global ethics and compliance program and officer, for example (as well as other relevant programs), and provides such programs the direct, visible support needed, working hard to create a culture where it is safe for employees to raise concerns and speak up, where performance management includes cultural metrics and where discipline, when necessary, is evenly meted out – to both executives and rank and file. There are a good number of companies – especially large, global, publically traded companies – that fit this bill.

The enlightened leader – This chief executive not only walks the talk, but also goes several steps beyond supporting internal and external ethics and corporate responsibility programs. He/she connects the dots between an ethical, responsible culture and ethical, responsible products and services. This chief executive creates the best internal structures, incentivizing and empowering employees to embed integrity not only into existing and new processes, but also into the company's products and services, and performance management.

FIGURE 10. Integrity leadership: The evolutionary ladder.

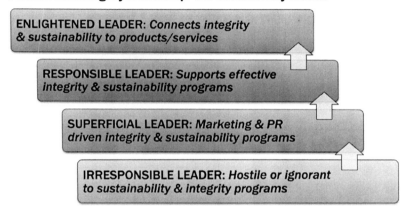

Very few companies have achieved this state of corporate 'nirvana' but there are a few like Unilever and Starbucks.

Second criterion: Integrity culture

There are five kinds of integrity culture that an organization can exhibit. These cultures are directly affected by the leader's integrity style and can best be gauged by looking at the intersection of two key factors: 1) the organization's integrity-related communications (governance, policies, statements and 'marketing'); and 2) the organization's integrity-related actions and behaviors, especially under challenging or crisis conditions.[33]

The key to understanding whether or not an entity has an 'effective' organizational culture of integrity is to understand the gap that may exist

– and the degree of such a gap – between the walk and the talk exhibited mainly by executive leadership (and to a certain extent the board) on issues of integrity within the organization.

FIGURE 11. Organizational integrity culture: A typology.

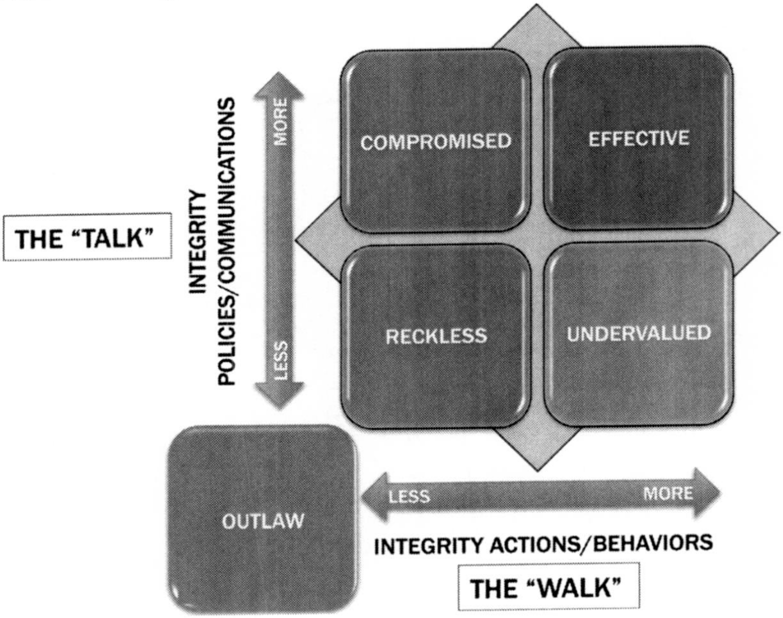

[Figure: A 2x2 matrix with axes. Vertical axis labeled "INTEGRITY POLICIES/COMMUNICATIONS" (THE "TALK") ranging from LESS to MORE. Horizontal axis labeled "INTEGRITY ACTIONS/BEHAVIORS" (THE "WALK") ranging from LESS to MORE. Quadrants: COMPROMISED (top-left), EFFECTIVE (top-right), RECKLESS (bottom-left), UNDERVALUED (bottom-right). An additional OUTLAW box appears at the far bottom-left outside the matrix.]

The key question is this one: does the organization's leadership integrity 'audio' (words) match its 'video' (actions)? The closer the match, the closer the actions are consistent with the communications, the greater the opportunity for an effective culture of organizational integrity and resilience. The farther apart, the more likely there is dysfunction or worse, possibly fraud, misrepresentation, recklessness or even criminality. The higher integrity leaders (the more evolved forms discussed earlier – the 'enlightened' and the 'responsible') will clearly be more closely associated with the more effective cultures (with the reverse being true as well). Let's take a look.

Based on these two criteria – the integrity 'walk' and the integrity 'talk' – we can distinguish five types of organizational integrity culture:

The outlaw organization – Is one in which criminal and/or civil violations are an integral part of doing business and there is no pretense on the part of leadership to be anything but an 'outlaw' organization. Examples include drug cartels, ISIS, Silk Road. The mission and vision of the organization is illegal or criminal by definition.

The reckless organization – Is one in which there are few if any deeds or words that pay lip service or otherwise to issues of good governance, risk and reputation management. Examples include Massey Mining, Anonymous. While the mission and vision isn't necessarily to flout the law, the overwhelming driver is results (usually financial) at all cost, regardless of the means.

The compromised organization – Is one in which few deeds (if any) match the more evolved form of marketing developed by management about the "deep commitment" of the organization to ethics and integrity and the 'zero tolerance' for deviations from integrity. Examples include BP's 'beyond petroleum' ad campaign prior to the Deepwater Horizon disaster.

The undervalued organization – Is one in which there is plenty of evidence of high integrity within the organization from its actions and behaviors but it doesn't do a good job of documenting or marketing this fact. The undervalued organization is a prime candidate for re-engineering to unlock trapped tangible and intangible value. Examples might include early stage companies with a great culture and excellent stakeholder relations which haven't been leveraged properly through effective internal and external marketing and communications programs.

The model organization – Is the one that has achieved a good balance between positive actions and behaviors and capturing such activities in its governance, policies, marketing and communications. The model organization has found a way in which to unlock tangible and intangible value, which could otherwise lie fallow within its organization. Examples are Starbucks, Chipotle, Fonterra.

What reputation risk management (RRM) strategy should your organization deploy?

Companies in the outlaw, reckless and compromised categories are at highest risk to their reputation. As long as they have reasonably effective leaders, entities in the model and undervalued categories will have the least imminent danger to their reputation because while they will not be free of possible reputational hits they will have processes, programs and crisis management in place to defend the resilience of the institution. That doesn't mean, however, that they cannot suffer from a swift and abrupt material hit to their reputations if, for example, an assumed exemplary (responsible or enlightened) leader is suddenly discovered to have done something unethical or illegal.

An entity with leadership that cares about integrity and long-term resilience is usually an entity that cares about other important metrics, performance, resilience, responsibility, sustainability, and holistic (financial and ESG) results for its stakeholders. The more a company thinks along these lines, the more it will be interested in reputation risk management – because it will see effective reputation risk management for what it is: a strategic asset, not a passing fad.

Leaders and directors that think this way, that is, understand that reputation risk is a strategic risk, also know that reputation risk can be a strategic opportunity. An opportunity for better products, processes, services, goods, culture, work environment. An opportunity for keeping stakeholders committed (customers), loyal (supply chain), happy (talented employees).

It is likely that a company with an un-evolved, incipient or non-existent integrity culture (i.e. the 'reckless' or 'endangered' categories discussed above) will have little or no interest in or ability to deploy a reputation risk management strategy. But there may be people within such entities (a board member or a senior executive) interested in developing greater organizational integrity and reputation management. To those brave souls we recommend a look at the tactical tools outlined in Chapter 5.

FIGURE 12. Reputation risk management (RRM) strategies most likely associated with each organizational integrity type.

ORGANIZATIONAL INTEGRITY TYPE	THE OUTLAW ORGANIZATION	THE RECKLESS ORGANIZATION	THE COMPROMISED ORGANIZATION	THE UNDERVALUED ORGANIZATION	THE MODEL ORGANIZATION
MOST LIKELY RRM STRATEGIES	• NO RRM • INCIPIENT RRM	• NO RRM • INCIPIENT RRM	• FACADE RRM • INCIPIENT RRM	• CHAOTIC RRM • EFFECTIVE RRM	• EFFECTIVE RRM • CHAOTIC RRM

Figure 12 conveys a couple of key concepts: first, that there are a couple of strategies that are possible within each type of organization – clearly, the 'model' organization is unlikely to operate with no RRM or 'incipient'

OPTIMIZING REPUTATION RISK MANAGEMENT:
THE NEXT STRATEGIC IMPERATIVE

RRM. Likewise, the 'outlaw' organization is unlikely to have 'effective' or even 'chaotic' RRM.

Most importantly, what this figure suggests is that the vast majority of organizations (which are likely to fall in the middle three categories) can do things to move up the RRM evolutionary ladder – they can do so by looking into and deploying many of the fifteen tools outlined in the toolkit in Chapter 5.

Visualizing reputation risk management within the organization: the big picture

Figure 13 places reputation risk management within the broader dynamics of what organizational integrity looks like. While this is a big topic that is beyond the confines of this Handbook, a brief discussion is merited to help place reputation risk management in its larger context. Here are a few key observations embedded in Figure 13:

1. Effective reputation risk management depends on the existence of proper *inputs* (programs like ethics, compliance, quality, etc.).

2. Effective reputation risk management depends on the existence of effective *enablers* (governance, leadership, risk management, strategy, incentives, etc.) that help to support and deploy the *inputs* or programs.

3. Reputation risk management should be part of strategic risk management which, in turn, is part of a key *enabler* – risk management or ERM.

4. *Inputs* and *enablers* together influence *outputs* – behaviors, actions, financial and other results.

5. When *outputs match or outperform stakeholder expectations* about an organization there will a *positive reputational gap* or an enhancement event with positive impacts on the organization and its value.

6. When *outputs under-perform stakeholder expectations* there will be a *negative reputational gap* with potentially damaging impacts on the organization and its value.

FIGURE 13. Reputation risk and the organization.

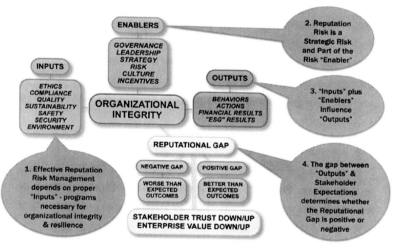

Beyond strategic: effective reputation risk management can be transformational

There is a promising next step for the development of effective reputation risk management: companies are beginning to demonstrate that they can

OPTIMIZING REPUTATION RISK MANAGEMENT: THE NEXT STRATEGIC IMPERATIVE

not only tame their risks generally and their reputational risk in particular but they can also transform those risks into business opportunities and even greater value. Enter Siemens and Fonterra.[34]

> **Siemens: Transforming systemic corruption into anti-corruption leadership**
>
> Between 2002 and 2006, Siemens, the 145-year-old German company, made bribery payments of about $300 million to facilitate the sale of power generation equipment in Italy, telecoms infrastructure in Nigeria and a national identity card project in Argentina. These turned into an international scandal as the German and US governments investigated, revealing that there had been a deep-seated culture of corruption at Siemens. Siemens eventually paid fines totaling $1.6 billion to both the US and German governments and close to another $1 billion in legal and related fees.
>
> The company changed both the supervisory and management boards, hiring a new CEO from outside Siemens for the first time in 145 years. The CEO hired a new General Counsel, replaced 80% of top-level executives, 70% of senior level executives and 40% of middle level executives. The company also conducted internal investigations, offered amnesty to employees who freely volunteered information on their roles, and fired those involved who had declined amnesty.
>
> Among additional proactive steps Siemens has taken to attack its previous culture of corruption are:

- Material strengthening of global internal ethics, compliance and anti-corruption programs – hiring the right people and adopting the right risk mitigation strategies.

- Complementary external actions, for example, Siemens Integrity Initiative with the World Bank.

- Commitment of $100 million to anti-corruption crusade in several countries including Egypt, Nigeria and South Africa.

Corruption is not something that is eliminated overnight or ever. But, by investing the proper resources and strategy into the creation of effective internal ethics, internal control and compliance programs and funding numerous external and highly visible anti-corruption initiatives, Siemens has turned a corner from being the corporate corruption pariah to a leader in the global fight against corruption with all the attendant reputational benefits and enhancements that go with this role including better relationships with key stakeholders – including, importantly, regulators and governments.

Fonterra: Transforming supply chain disaster into direct ownership revenues

New Zealand dairy company Fonterra Cooperative Group Ltd faced a crisis when their Chinese manufacturing partner, Sanlu Group Co Ltd, produced and sold milk powder contaminated with poisonous melamine, killing six babies and sickening approximately 290,000 others. When Sanlu was forced to file for bankruptcy, Fonterra was able to maintain its distance and independence. Learning from

OPTIMIZING REPUTATION RISK MANAGEMENT: THE NEXT STRATEGIC IMPERATIVE

> this scandal, Fonterra shifted its business model to manufacture and sell milk powder directly to Chinese consumers by eliminating its supply chain model and replacing it with a direct ownership model. Fonterra increased sales in China by both importing dairy products and also producing these products with local farms in China. The company was therefore able to increase its market share and profits.
>
> In August 2013, Fonterra faced another scandal when it was believed that bacteria causing botulism had contaminated its milk supply. The Chinese government stopped all imports of milk powder from New Zealand and Australia. Learning from its past experience, Fonterra acted quickly to contain the situation, address public concerns and restore confidence. Today, Fonterra is making record revenues and profits in China.

These are but two examples of companies transforming a serious strategic and reputational risk into an opportunity for better business. These are not unique cases – this is something other smart organizations can emulate. All they need to do is understand their reputation risk, triangulate it properly and deploy an optimized strategy that not only looks at mitigating the risk but transforming it into stakeholder value.

Conclusion: The Way Forward

THIS HANDBOOK has attempted to shine a light, provide an understanding, perhaps a lens, into how managers, executives and boards can gauge, develop, implement and oversee effective reputation risk management within an organization – whatever shape, size, purpose it may have.

Not every organization is going to care about having effective reputation risk management – not because they don't need it (everyone does) but because many are not even aware of reputation risk as a strategic and material issue, let alone how to begin to tackle it. . . until they are forced to.

In this Handbook, we have demonstrated that reputation risk management is a strategic must for any organization in today's age of hyper-transparency – sooner or later you and the entities you are associated with will need to pay attention. It can be done the responsible, anticipatory and preventative way – by gauging current conditions and building effective preventative measures. Or it can be done the irresponsible, chaotic and hard way by waiting for the inevitable reputational crisis to hit and reacting, in a disadvantaged and sub-optimal way, after the fact.

Smart organizations are prepared to, and will deal with, reputational hits effectively when they happen; they will also go the extra mile to delve into what happened and how to do things better in the future often even questioning their very business model.

CONCLUSION: THE WAY FORWARD

We are at the very beginning of understanding all the ins and outs of reputation risk but it is clear that it is here to stay and with continued rapid technological and communications change it will only grow in size and complexity. It's time to start somewhere and this Handbook is designed to give practitioners some useful tools, perspectives and strategies to get started in tackling this important subject.

Notes

1. Socrates: http://www.goodreads.com/quotes/365753-regard-your-good-name-as-the-richest-jewel-you-can; Syrus: http://en.wikiquote.org/wiki/Publilius_Syrus; Lincoln: http://www.goodreads.com/quotes/28413-character-is-like-a-tree-and-reputation-its-shadow-the; Buffet: http://www.goodreads.com/quotes/148174-it-takes-20-years-to-build-a-reputation-and-five; and Buffet: http://blogs.wsj.com/deals/2011/03/31/warren-buffett-on-ethics-we-cant-afford-to-lose-reputation/

2. *Source*: Bonime-Blanc, A. 2013. How to prevent the next business scandal. *Ethical Corporation Magazine*, 18 December, http://www.ethicalcorp.com/business-strategy/globalethicist---how-prevent-next-business-scandal

3. Edelman Trust Barometer, 2014. http://www.edelman.com/insights/intellectual-property/2014-edelman-trust-barometer/; Bonime-Blanc, A. 2014. Do banks care about reputation risk? *Ethical Corporation Magazine*, 5 March, http://www.ethicalcorp.com/business-strategy/globalethicist---do-banks-care-about-reputation-risk

4. Valukas, A. 2014. Report to Board of Directors of General Motors Company regarding ignition switch recalls, 29 May, http://www.nytimes.com/interactive/2014/06/05/business/06gm-report-doc.html?_r=0

5. http://www.nytimes.com/2012/04/22/business/at-wal-mart-in-mexico-a-bribe-inquiry-silenced.html?pagewanted=all&_r=0

6. http://www.bloomberg.com/news/2011-07-12/news-corp-s-lost-7-billion-shows-investor-concern-over-hacking-fallout.html

7. *Sources*: Deloitte. 2013. Strategic Risk Survey 2013, http://deloitte.wsj.com/riskandcompliance/files/2013/10/strategic_risk_survey.pdf; Allianz, 2014.

NOTES

Risk Barometer 2014, http://www.agcs.allianz.com/assets/PDFs/Reports/Allianz-Risk-Barometer-2014_EN.pdf; Eisner Amper, 2014. Concerns about risks confronting boards, http://www.eisneramper.com/uploadedFiles/Resource_Center/Articles/Articles/Concerns-Risks-Survey-2013.pdf; Clifford Chance and the Economist Intelligence Unit, 2014. View from the top: A board-level perspective on current business risks.

8. http://www.ft.com/intl/cms/s/0/77574c74-1377-11e4-84b7-00144feabdc0.html

9. There are several good resources that go beyond the typical public relations treatment of reputation including: 1) two books by Dr Nir Kossovsky: 2012. *Reputation, Stock Price, and You: Why the Market Rewards Some Companies and Punishes Others* (New York: Apress), and 2010. *Mission Intangible: Managing Risk and Reputation to Create Enterprise Value* (Bloomington, IN: Trafford); 2) work by the Reputation Institute (**http://www.reputationinstitute.com**); 3) work by Corporate Excellence: Centre for Reputation Leadership (**http://www.corporateexcellence.org**) including the 'Reputational Risk Management Model' (2014), and Carreras, E., Alloza, A. & Carreras, A. 2013. *Corporate Reputation* (Madrid: LID Editorial); 4) the articles and publications of Jorge Cachinero and Llorente & Cuenca some of which are available here: **http://www.llorenteycuenca.com/index.php?id=9**; and 5) **Garcia, M. 2011. Riesgo Reputacional. Auditoria Interna 94, Instituto** de Auditores Internos de España.

10. Alloza et al., *Corporate Reputation*.

11. http://www.reprisk.com and http://www.reputationinstitute.com.

12. http://www.rmmagazine.com/2014/04/01/how-to-manage-reputation-risk/

13. Economist Intelligence Unit. 2013. A crisis of culture: Valuing ethics and knowledge in financial services.

14. Eccles, R.G., Ioannou, I. and Serafeim, G. 2012. The impact of a corporate culture of sustainability on corporate behavior and performance. Working

Paper, May, http://www.natcapsolutions.org/events/2014/Benedictine University/HBR_The-Impact-of-a-Corporate-Culture-of-Sustainability.pdf

15. Kossovsky, *Reputation, Stock Price, and You*.

16. Bonime-Blanc, A. 2014. Risk and opportunity: The role of stakeholder trust. *Ethical Corporation Magazine*, 8 May, http://www.ethicalcorp.com/stakeholder-engagement/globalethicist-risk-and-opportunity-role-stakeholder-trust

17. *Oxford English Dictionary* definition of 'risk' from c. 1655 (first known definition), http://en.wikipedia.org/wiki/Risk

18. ISO 31000 (2009), http://www.iso.org/iso/home/standards/iso31000.htm

19. http://www.oxforddictionaries.com/us/definition/american_english/reputation.

20. http://www.merriam-webster.com/dictionary/reputation.

21. Cachinero, J. forthcoming. Reputational risks.

22. http://abcnews.go.com/International/justine-sacco-fired-tweet-aids-africa-issues-apology/story?id=21301833.

23. *Sources*: http://www.nytimes.com/2014/03/22/business/fallout-from-snowden-hurting-bottom-line-of-tech-companies.html?hp&_r=0; Bonime-Blanc, 2014. Risk and opportunity.

24. http://blogs.wsj.com/moneybeat/2014/06/26/thai-food-companies-hit-by-u-s-human-trafficking-downgrade/

25. Kossovsky, *Reputation, Stock Price, and You* and *Mission Intangible*.

26. See sources listed in note 6.

27. *Source*: Bonime-Blanc, A. 2014. The biggest risks nobody talks about. *Ethical Corporation Magazine*, 7 February, http://www.ethicalcorp.com/business-strategy/globalethicist---biggest-risks-nobody-talks-about

NOTES

28. *Source*: Bonime-Blanc, A. 2013. Risky business. *Ethical Corporation Magazine*, 1 October, http://www.ethicalcorp.com/business-strategy/globalethicist---risky-business

29. Cachinero, J. 2011. Reputation is here at last. Llorente & Cuenca, June. Bonime-Blanc, A. 2013. Who runs the reputational risks? *Ethical Corporation Magazine*, 30 April, http://www.ethicalcorp.com/business-strategy/globalethicist---who-runs-reputational-risks; Bonime-Blanc, A. 2014. Implementing a holistic governance, risk and reputation strategy for multinationals. *Ethical Boardroom*, 1 September, http://ethicalboardroom.com/risk/implementing-holistic-governance-risk-reputation-strategy-multinationals-guidelines-boards/

30. *Source*: Bonime-Blanc, A. 2014. The four maxims of board reputation risk oversight. In B. Kimmel (ed.) *Trust Inc.: A Guide for Boards and C-Suites* (Dallas, TX: Next Decade, Inc.).

31. The two retail industry worker safety initiatives and an auto industry supply chain ethics initiatives are available here: **http://www.bangladeshaccord.org/2014/04/the-one-year-anniversary-of-the-rana-plaza-building-collapse/**; http://www.bangladeshworkersafety.org/about/about-the-alliance; http://www.aiag.org/staticcontent/files/CorporateResponsibility GuidanceStatements.pdf

32. This discussion is based on the author's research and article: Bonime-Blanc, A. 2013. Take it from the top. *Ethical Corporation Magazine*, June, **http://www.ethicalcorp.com/business-strategy/globalethicist---take-it-top**

33. This discussion is based on unpublished research and proprietary methodology developed by the author.

34. Löscher, P. 2011. The CEO of Siemens on using a scandal to drive change. *Harvard Business Review*, November, **http://hbr.org/2012/11/the-ceo-of-siemens-on-using-a-scandal-to-drive-change/ar/** (accessed 26 November 2013); NPR. 2012. Siemens changes its culture: No more bribes. *NPR.org*,

THE REPUTATION RISK HANDBOOK:
SURVIVING AND THRIVING IN THE AGE OF HYPER-TRANSPARENCY

1 May (accessed 26 November 2013). The Fonterra and Siemens cases are based on unpublished and ongoing research conducted by the author, with the assistance of a Columbia University, School of International & Public Affairs, Masters in International Affairs students since 2013.

Lightning Source UK Ltd.
Milton Keynes UK
UKOW02f2155120115

244381UK00004B/377/P

9 781910 174302